SEDIMENTOGRAPHICA

Photographic Atlas of
Sedimentary Structures

SEDIMENTOGRAPHICA

Photographic Atlas of
Sedimentary Structures

second edition

Franco Ricci Lucchi

NEW YORK   COLUMBIA UNIVERSITY PRESS

Columbia University Press
New York   Chichester, West Sussex
English edition: copyright © 1995 Columbia University Press
Italian edition: copyright © 1970, 1992 Zanichelli S.p.A.
Bologna

Library of Congress Cataloging-in-Publication Data

Ricci Lucchi, Franco.
[Sedimentografica. English]
Sedimentographica : photographic atlas of sedimentary
structures / Franco Ricci Lucchi.—2d ed.
p.   cm.
Includes bibliographical references and index.
ISBN 0-231-10018-3 (acid-free)
1. Sedimentary structures—Pictorial works.   I. Title.
QE472.R5313   1995
552′.5′0222—dc20   94-24952
CIP

♾

Photographs with no indication of source belong to the author.

Drawings by Gianni Cavalcoli and Enrico Craici.

Printed in the United States of America

c 10 9 8 7 6 5 4 3 2 1
p 10 9 8 7 6 5 4 3 2 1

10759685

ℛᵉᵒ

Contents

Preface

This photographic atlas illustrates the structures and other geometric aspects of sediments and sedimentary rocks. It has been translated into English from the 1992 second Italian edition, with some modifications and additions. The first edition, dating back to 1970, was directed mainly to students graduating in earth sciences; it was more technical than the present one and required some previous knowledge of sedimentology and geology. The second edition has been written in a more accessible style and is addressed to students approaching sedimentology and sedimentary geology for the first time. It should also pique the interest of amateur scientists and people who are curious about natural objects. These specific objects, i.e., sedimentary strata and their features, are visible to the naked eye in several places: mountainous areas, river banks, seashores, quarries, and road cuts. A book like this can thus be used as a sort of guide to a large, open-air museum to be visited on foot or, in many cases, also by car. I wish to point out, however, that it is not a complete catalog of these phenomena.

In sediments and sedimentary rocks, structures represent what lineaments are to a face; they reveal the sediment or rock character and reflect its origin. Their shape often has the elegance and the grace of sculpture, painting, and tapestry.

Many other aspects of sediments—chemical and mineralogical composition, texture, physical, and technical properties—can be observed or measured. All of them are useful for identifying and characterizing sedimentary products. They have been covered in other books and manuals on sedimentology, both general and specialized, but a previous knowledge of their content is not necessary for reading and consulting this book. Here, basic concepts and processes are introduced along with the description and presentation of sedimentary structures; this is done in the simplest possible way, which obviously cannot satisfy a reader who demands every detail. The point is that you do not need to read other books on sedimentology or geology before enjoying this one. In this sense, *Sedimentographica,* although dealing with only one aspect of sediments and consequently being complementary to other works, can claim a certain autonomy: you can approach it with the support of the essential scientific basics learned in high school. Frequent analogies are made with objects and phenomena familiar to everyday life to ease understanding of processes and mechanisms; many cross-references also help the reader to compare, contrast, and discriminate among structures.

All kinds of structures, both primary and secondary, are dealt with in this atlas. They are grouped into broad genetic categories; not every specific type, however, is illustrated. The main purpose is not to show all known structures but a selection of the most significant ones, reflecting the more basic and representative processes that act on sedimentary materials: water currents, waves, glaciers, wind, tides, gravity, living organisms.

I cannot avoid a certain bias toward examples and areas more familiar to me, but I have tried to remedy this as much as possible by using pictures taken during excursions to various parts of the world or asking colleagues for them. Approximately 70% of the images are from Italian environs, 30% from the rest of the world. Italy is famous for its vast collection of manmade monuments and art treasures, less for its natural rocks and geology (except, perhaps, for the active volcanoes and the Serapes market with its perforated columns, immortalized by Lyell's treatise). The book will show that this small area at the center of the Mediterranean is also a repository of geological treasures. In particular, the mountainous backbone of the Italian peninsula boasts many superb exposures of ancient carbonates, evaporites, and deep-water deposits, including classical examples of turbidites.

The photographic documentation of sedimentary structures is immense: at least 1,500 references can be quoted for the last twenty years. The main inconvenience is that most of this iconographic material is scattered among too many papers, generally but not always available in university libraries: only a few general articles and textbooks exist (see References). Among them, the atlas of Pettijohn and Potter (1964) stands out as a milestone, and has greatly influenced the original project of this book.

Providing a tool for geological curricula and fieldwork is the main purpose of an atlas, but not the only one; not the least of my intentions, especially in preparing a new edition, was to attract the curiosity of the layperson and to make the public at large more aware of earth science phenomena. Sedimentary structures, after all, are out there to be used not only for technical or scientific purposes, but simply to be looked at; they are "forms that show themselves to the world" (A. Portmann 1969).

The book is dedicated to the memory of a person who encouraged me and made the first edition possible—Delfino Insolera, whom friends remember for his curiosity about nature, his love of science as a human achievement, and his great ability in communicating his enthusiasm to other people.

I wish to thank all the colleagues and students who kindly provided pictures and suggestions. In particular, I want to express my gratitude to Guido Piacentini and Vanna Rossi, two photographers gifted with a bright and delicate touch, who provided many fine images. My thanks also to Paolo Ferrieri, who printed the 2,000 pictures that served as a base for the final selection; to Maria Angela Bassetti, who helped prepare the pictures, the glossary, and the index; to my wife, Nevia, who read the text and improved the language; and to researchers and technicians of the Institute for Marine Geology of C.N.R. in Bologna, who opened their archive of cores and seismic records to me.

My last thought, however, is for several generations of students, who suffered before, during, and after examinations, and thus helped to reveal deficiencies, limits, and inconsistencies of the first edition.

Introduction

How to Use the Atlas

With the exception of this introductory section, the pictures in the atlas are labeled and numbered as plates. Each plate consists of one or more pictures related to the same subject, which gives it its title. A running head at the top of page recalls the type, or family, to which the example shown is assigned (e.g., erosional structures).

Types are defined in a broad way, and individual structures are not further classified or listed following a systematic order; rather, they are ordered in sequences that outline "routes" for reading. Starting with the basic or typical features, some variants are then shown to point out similarities and differences. Certain sequences zoom in the same object, or phenomenon, as in the case of stratification (from a wide panoramic view to a close-up of individual beds). In essence, an attempt is made here to interweave the threads of a series of topics, whereas the previous edition was organized more in the style of a deck of cards.

Because sedimentary features are found in a great variety of sizes, geologists usually place, within the picture frame, persons or objects to give an idea of the actual size or *scale*. For an average distance (middle scale), a geologist hammer is commonly used. in close-ups and hand specimens (small scale), a variety of objects, from a meter to a lens cap, a coin or a paper clip, may be used.

For readers to get a correct spatial perception of the structures shown in the photographs, which were shot from a variety of distances and visual angles, mostly in the field but also in the lab, something must be said about *what* the subjects are and *how* they show up:

• in *Modern* sediments, just deposited and mantling a topographic surface exposed to the atmosphere or under water (see figure 1), structures appear as morphological details of that surface. This, however, is only a partial expression of structures, which are three-dimensional objects, as you can see by cutting a section, or a trench, across the depositional surface (figure 2). The sediment can be impregnated with resins and emulsions that consolidate it and allow coherent slices to be picked up; if the liquid simply adheres to the surface of the grains exposed along the section, a contact replica (a peel) can be obtained (figure 3). Sedimentary structures can thus be taken to the lab and carefully examined in proper light conditions, or put into exhibits;

• in *Recent* sediments, immediately underlying the Modern ones and having a maximum age of 10,000 years (the conventional time interval called Holocene, the youngest part of the Quaternary), structures can be observed (only) in stratigraphic sections, and are part of the stratigraphic description. Sections can be natural (e.g., river cuts and terraces) or artificial (road-cuts, quarries). Their sizes are quite variable, both vertically and laterally: the narrowest can be seen in cores, cylinders of sediment or rock that are extracted from buried sediments by various drilling and sampling devices penetrating the sediment. Sediment cores can be 2 or more meters long but only

Figure 1. Ripples on the sandy surface of a beach at low tide. These small-scale structures were formed by waves when the beach surface was under water (upper part of photo), and subsequently modified by the low tide. The ebb flow remolded the sand and made its own ripples, which have a different shape (asymmetrical) and orientation (from lower right to upper left) with respect to ripples produced by waves. Part of the surface was completely smoothed by shallowing water (upper right corner). Ripples are formed and canceled many times in a day or in a week; sometimes, but rarely, they can fossilize. Imagine that the sea level suddenly rises, submerging the beach surface to a depth out of reach for waves. The ripples will then be covered by mud and preserved. PHOTO G. C. PAREA 1970.

Figure 2. Lingering in the same place (more exactly, in the part of the beach that is normally emerged) and cutting a vertical section in the sand with a knife, one can observe the profile of several surfaces produced by sedimentation or erosion in recent times. Some of these surfaces are flat, others are wavy; the former reflect the smoothing effect of waves, the latter record partly preserved ripples. Darker and lighter bands indicate depositional laminae and sets of laminae (organic matter, iron minerals, or other peculiar grains are concentrated in dark laminae); their superposition underlines the vertical accretion of the beach.

The change in geometry of the buried surfaces suggest that the shoreline shifted a little during the time recorded by sedimentation, in such a way that the spot we are in remained alternately in submerged and emerged conditions. By comparing this photo with that of figure 1, you can see that the *interfacial* expression of ripples is much more evident than their *cross-sectional* view. Nevertheless, ripples *can* be recognized in section and this is important because scientists usually work on sections when studying Ancient sediments. PHOTO G. C. PAREA 1970.

some 10 cm wide, and represent an almost linear, vertical record of sedimentation (figure 4);
• in *Ancient* sediments, lithified to a various extent and transformed into rocks, things can be more complicated. Surfaces exposed by erosion, landslides or human activity, form *outcrops* and may range from almost plane and even to irregular and rugged; it depends on various factors, such as the type of sediment or rock (lithology), the type and duration of weathering, the orientation of the surface with respect to the wind and sun, the erosional agent (wind, water, ice), and the chemical or mechanical action of organisms (plant roots, lichens, etc.).

Many outcrops are markedly bidimensional and can be called *cross-sections*. Sedimentary rocks can be recog-

nized there because of their stratified, or bedded, character; visible lines represent the intersections between bedding planes and the plane of section. Stratification is not always apparent, and many outcrops look unbedded or poorly bedded. Except in arid zones, devoid of vegetation, outcrops are generally discontinuous and separated by areas where the rock is covered (by soil, detritus, plants, buildings, roads).

Bedding can be deformed or crushed by tectonic forces operating in the outer, rigid earth shell (crust and lithosphere) or in the deeper mantle; in the case of mild deformation, strata are simply tilted, or rotated around an horizontal axis, and appear inclined with respect to their original horizontal setting (figure 5). It is, in fact, assumed as a rule that depositional surfaces are flat and horizontal. A

Figure 3. Mold of a section to that of figure 2, dug in Recent sands of a Dutch tidal flat (North Sea). The laminated sand has been impregnated with resin, which has been taken away after drying. The structure shown is cross-lamination, changing in scale from the lower to the upper half. Small-scale cross-laminae reveal the formation and migration of ripples, while medium-scale laminae reflect similar but larger structures (subaqueous dunes). PHOTO A. COLELLA 1985.

Figure 4. Sediment cores collected in the Red Sea have been cut longitudinally to show the stratigraphy; the top of this composite section is to the upper right. Note the change from well-laminated (or thinly bedded) sediment in the upper portion to relatively homogeneous, structureless sediment. All these sediments are fine grained, and the bedding is caused by a different mechanism as compared with previous examples (fallout from a water column instead of lateral traction on the bottom). The dark layers have a high content in organic matter.

Cores are cylindrical, vertical samples wrapped by a plastic liner. Various types of sediment corers are in use; light corers (piston, gravity, etc.) are employed to sample surficial sediments of the sea bottom. They penetrate for some meters after being dropped from a boat. Hard rock cores are obtained by rotary drilling equipment that cross geologic formations hundreds of meters thick. PHOTO COURTESY OF INSTITUTE FOR MARINE GEOLOGY C.N.R. 1983.

secondary inclination is, however, almost the norm for bedding as some tectonic stresses affect, or affected to a various degree in the past, every part of the earth's surface. In certain places, sediments are laid down on sloping surfaces, and the beds have a *primary inclination.* Secondary inclination must not be confused with the original one, which is also called clinostratification, or clinoforms.

This tells us that the study of Ancient sediments cannot be done apart from fundamentals of geology: sedimentary rocks are just a type of rocks, which can have been involved in various geologic processes, and experienced many vicissitudes after their formation.

A knowledge of the principles and rules of stratigraphy is especially important for correctly reading and interpreting sedimentary phenomena. Beside the assumption of *original horizontality* of strata, the principle of *superposition* (beds at the bottom of a pile or succession are older than those at the top) and that of *intersection* (features and objects that cross or truncate others are younger: see, for instance, fractures versus bedding in figure 6) must be taken into account. Moreover, there are *way-up* criteria to establish whether beds are in normal position or upside down (figure 5) because of tectonic movements and rotations. Every time we geologists observe inclined strata, we should ask ourselves: are these beds upright or overturned? In other words: are they inclined less or more than 90° from their original position? The answer can be provided, in many cases, by sedimentary structures, if they are present within the beds or on their surfaces (bedding planes). If we are able to identify the structures, their shape and orientation give us the vertical orientation, or *stratigraphic polarity,* of the beds.

The way-up criterion works like this: in the normal position, the lower bedding plane, or base of a bed, faces downward, no matter how much it is inclined; if the position is inverted, the base faces upward. It is thus fundamental to determine whether an exposed bedding

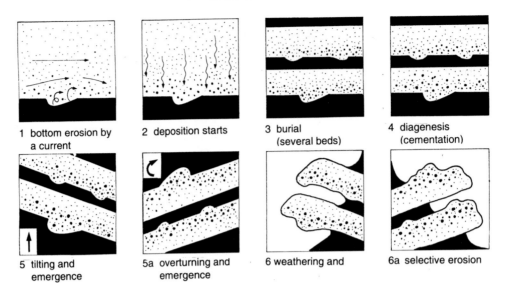

Figure 5. How sole marks are formed and preserved (as molds). They can be used for way-up recognition in stratigraphy: 1. bottom erosion by a current; 2. deposition starts; 3. burial (several beds); 4. diagenesis (cementation); 5. tilting and emergence: 5a. overturning and emergence; 6, 6a. weathering and selective erosion.   FROM RICCI LUCCHI 1970.

plane is the base or the top of a bed. If you look again at figure 5, and compare frames 5 and 5a (or 6 and 6a), this point will be made clear. Notice also, in the same example, that you are not observing the original structures, but molds or casts of them: depressions formed at the top of a bed appear as relieves at the base of the overlying bed, and vice versa. What is preserved is often the filling of a scour, not the scour itself. This depends on the different behavior of sediments (for example, sands interbedded with muds), when they are compacted, cemented and, later on, weathered and eroded. Remember, then, that it would be wrong to infer the type of structure from its present position and the orientation of the bed in which it is found. The right procedure is just the opposite: the original morphology and orientation of the structure must be recognized first, then the bed orientation can be inferred.

To understand a structure, one should see where, when and how it is formed in Modern environments, or try to reproduce it with experiments in controlled conditions (laboratories, flumes, wind tunnels, and so on). The implication is that one relies on the fundamental principle of uniformity, or *actualism* ("the present is the key to the past"), which is a guide for all branches of geology. That

Figure 6. A sandstone bed in vertical position. The photo has been taken from above (it crops out on a littoral terrace); the original top is recognizable by the finer grain size and the presence of laminae. See network of fractures (secondary structures) almost at right angle with bedding (primary structure).

is not an absolute rule, and should not be applied too rigidly. Processes active today were certainly active in the geological past, but their intensity or duration may have changed. When the vegetation did not exist on land surfaces, the rate of erosion was higher because the bare ground had no protection from wind or flowing water. When the atmosphere contained little or no oxygen, the chemical alteration of rocks and the decomposition of organic matter, which influence the rock cycle and sedimentary processes, were different. With the evolution of the biosphere, the interactions between organisms and abiotic components of the environment, including sediments, became more and more intricate.

Other examples could be quoted, but these are sufficient to show that the rate of change in earth systems has not been uniform, and that their present condition is the result of their evolution in time. In some respects, science says that *the past is the key to the present*. It can be said geologists, endowed with imagination, ingenuity and spirit of observation, were able to infer, from the rocks themselves, processes that were detected or proven only later: catastrophic floods due to collapsing of massive ice dams (e.g., Lake Missoula in the western United States), desiccation of an entire sea (the Mediterranean at the end of Miocene), and turbidity currents and large submarine slides (from the observation of bedding and structures in sandstones of the Alpine fold belts).

Thus, we geologists are all uniformitarian if that means that "past geological events can be explained by phenomena and forces observable today" (AGI Glossary, 1987), as the basic laws of nature have not changed since the formation of our planet. We realize, however, that: 1) specific processes, which are a particular expression of these laws, have changed, especially in the biological and biochemical domains; 2) a certain dose of "catastrophism," the old archrival of actualism, must be admitted as far as natural phenomena are concerned.

The last point implies that actualism does not mean "every day" or "every year" events only; there are so-called rare events occurring sporadically, without a regular periodicity, which deliver enormous amounts of energy over restricted areas and during short-time intervals. These events, which are also called "catastrophic," recur over time spans that can exceed a human life span, several generations, or the whole human history. Large fluvial floods, tsunami waves, earthquakes of large magnitude, and exceptional volcanic eruptions are all cases in point. Some of these had a tremendous impact on early civilizations, and originated legends and myths such as the submergence of Atlantis or the Great Deluge. The human mind tends to record these natural events as unique and to charge them with meanings and purposes known only to God.

Modern statistics tell us that the rarer the events, the bigger they are; in other words, there is an inverse relationship between their magnitude and their frequency (Poisson's rule). That is why no human beings witnessed the collision of a large meteorite or comet body (but dinosaurs did, 65 million years ago). If the average recurrence time of such an event is, for example, 10 or 20 million years, should science say that it represents an actualistic phenomenon or not? By applying the concept too literally, one would be induced to consider as actualistic only the scattered rain of micrometeorites that fall continuously on Earth (see figure 13). On the other hand, there is a probability, however small, that a celestial body of considerable size will impact on our planet this year or next year: several near misses have already been reported in the last decades. It is, therefore, more reasonable to encompass all *natural* events and processes in a uniformitarian view of the world.

After this long digression occasioned by the problems of interpretation of Ancient sediments, I come back to more practical items by considering the last two points:

• several structures of intermediate (meso) or small scale were photographed on isolated blocks of rock, which detached from cliffs or quarry walls; their orientation, in that case, has no stratigraphic value, but can be the right one for taking a good photograph (appropriate light incidence, etc.) and emphasizing morphological details; look, for example, at plates 159 to 163;
• large-scale structures are rarely exposed in their entirety in outcrops; for a better understanding of their geometry, some examples of seismic profiles have been used (they show buried beds: see more in the following section and plates 1 and 2); when possible, a seismic section and a natural section have been put side by side, or in sequence, to make comparisons easier.

Sedimentary Structures: Preliminary Remarks

Deciphering the visible marks left by natural processes on rocks and sediments is a fascinating part of a geologist's work; physical processes such as currents and waves are responsible for most of these structures, but chemical reactions and biological activity also contribute. So, even where body fossils of organisms (shells, bones, plant remains) are lacking, the traces of their past existence can be preserved in sediments as a disturbance caused by their passage, metabolism, rest, or rooting. All these marks and traces can be encompassed under the phrase *sedimentary structure;* they consist of objects and forms that are produced by sedimentary processes and are preserved in rocks. That structures can fossilize or, in other words, have a preservation potential, make them interesting to geologists. Some structures are very decorative and can attract the attention of a passer-by, who perhaps attributes them to some prank of nature. Even among scientists, not many years ago many structures were called "hieroglyphs."

Apart from some ingenious intuitions expressed by Leonardo da Vinci, serious attempts to understand the meaning of structures were not undertaken until the nineteenth century when eminent geologists like C. Lyell,

H. C. Sorby, and G. K. Gilbert realized that these morphological features of sediments and sedimentary rocks could offer clues to their origin. This in turn could contribute to finding useful materials and resources associated with sediments, such as coal.

A great impetus to the study of sediments and the founding of sedimentology as a special branch of earth sciences, came from the growing importance, after 1930, of the oil industry and mass motorization. Hydrocarbons are intimately associated with both our life and sediments: their source is a sediment rich in organic matter, and their ultimate repository, or reservoir, is made of porous sedimentary bodies. Structures are useful for understanding the environment and the tectonic setting in which sediments are deposited and transformed into rocks, coal, or oil. Sedimentary processes are responsible for deposition, diagenetic processes for *post-depositional* transformations.

To make the best use of structures as a tool, both their potentialities and their limitations must be known. Among the many structures produced in sediments, only a few are bound to preservation; most of them are canceled by the same sedimentary processes, or by prolonged exposure to weathering and erosion. The surface of a beach, for example, is a sort of palimpsest (figures 1 and 2): in the emerged part, ripples produced by wind or the tracks of our feet are flattened by high tides or cleaners; in the submerged parts, waves continuously remold the sandy bottom. In general, a newly deposited sediment is subject to *remobilization* by various agents, especially by cataclysmic events such as storms or gravity slides. Only when it is buried under other sediments, can it preserve its structures. How many and which structures will be found in the stratigraphic record depends not on a single event but on the whole history of the sediment, including diagenetic changes (diagenesis sometimes enhances the structures but in other cases efface them). This history can be synthesized in the following steps: *primary deposition, remobilization, redeposition, burial.* The loop remobilization-resedimentation can be repeated many times before arriving at the definite emplacement. When the loop is not activated, and deposition is immediately followed by burial, the chances of preservation of primary structures are maximized.

After burial, tectonic processes can add to diagenesis in modifying sediments and their structures, which react to tectonic stresses with various types of deformation. A *brittle* behavior is typical of an indurated, rigid sediment; it is dismembered into blocks or slices by ruptures and faults: within the blocks, no deformation occurs. A *ductile* behavior, often but improperly called "plastic," means that the sediment is still relatively soft (because of its fine grain size and water content) or was mollified by high temperatures and pressures. Folds and shear planes represent the deformation in this case. Shear planes can be discrete and far apart or closely spaced; in the latter case, *cleavage* is produced in the rock. Cleavage and small-scale folding are an example of pervasive deformation; an analog for brittle materials is given by comminute fragmentation, or *brecciation.* In pervasively deformed sediments, sedimentary structures can be completely obliterated and replaced by *tectonic structures.* In certain cases, small- and medium-scale tectonic structures mimic the sedimentary ones, and attention must be paid to avoid confusion (see, for instance, plates 177 and 178).

The static pressure due to the weight of overlying materials contributes to the *physical aspect of diagenesis* in buried sediments: water is squeezed out from pores, the solid particles come closer, and the sediment volume decreases. Thus the sediment is compacted; three-dimensional structures are flattened and squashed because of compaction. The chemical aspects of diagenesis include exchanges of matter and energy between sediment particles and interstitial fluids: the result may be dissolution of some materials, changes of composition, formation of new minerals, hydration or dehydration, cementation by salts deposited in pores and cavities. Overall, diagenetic processes tend to lithify a sediment, or transform it into a rock (somebody speaks of lithogenesis). Buried sediments can also be affected by heat sources, such as magma bodies, and be "cooked" or metamorphosed. The boundary between diagenesis, as part of the sedimentary or surficial processes, and metamorphism, is often subtle because temperature and pressure increase gradually with depth in the subsurface.

Not all of diagenesis is done underground, however. A fall in sea level or a crustal uplift may bring marine sediments, still uncompacted or unlithified, in the emerged domain, where erosional processes predominate and can exhume them. Materials escaping erosion may then undergo a subaerial form of diagenesis, consisting of dissolution and/or cementation. This *vadose* diagenesis is caused by drastic chemical changes in the sediment pore fluids, when meteoric water (i.e., slightly modified rainwater) replaces saltwater, and is transitional to soil formation (pedogenesis).

Bedding surfaces represent discontinuities in sedimentation; before diagenesis, they are not so obvious and the beds stick together; the effect of diagenetic processes (compaction, differential cementation, etc.) is to enhance the discontinuities as contrasts in physical properties. Beds can thus be separated by both natural and artificial causes; moreover, they send a more distinct signal when penetrated by acoustic waves. In that case, buried beds are (seismic) *reflectors.*

Several structures may be found on exposed bedding surfaces, which is no surprise because all these surfaces were, before burial, part of the bottoms of rivers, lakes, and seas (figure 7). In essence, they were *depositional interfaces,* either subaerial (between air and sediment) or subaqueous (between water and sediment).

Structures present on bedding planes are also visible as lines in sections and trenches; the same object can consequently show us an *interfacial* or a transfacial

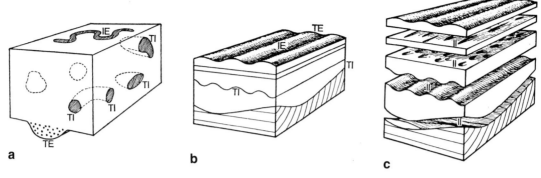

Figure 7. Orthogonal sections of beds to show different views of sedimentary structures. In **a,** traces and burrows of organisms and a basal scour fill; in **b** and **c,** ripples, cross-laminae (foreset), and sole marks.

(*cross-sectional*) view. Remember that both are an expression of the same object. Ripple marks are among the most typical interfacial structures at the top of beds, as shown by figures 1 and 2; structures and traces occurring at the base of hard beds (generally as molds) are collectively called *sole marks,* or basal structures, and may show a profusion of shapes.

Other depositional surfaces can form inside the beds when sedimentation is not continuous but occurs in steps or pulses; laminae can thus be recognized. A bed can be entirely laminated or only in part; in the latter case, laminated is distinguished from nonlaminated, or *structureless,* portions. These portions are called intervals, horizons, or divisions. The effects of diagenesis are felt also by all surfaces internal to beds (*intrastratal*), which become surfaces of weakness; laminated portions can thus be detached from the others, or split into individual laminae and sets of laminae. The rock is called fissile, or flaggy. Minor sedimentary structures can be observed on intrastratal surfaces, or *parting planes.*

Finally, diagenesis can produce structures of its own within sediments, e.g., concretions, nodules, cavity linings, or fillings. These structures are called secondary, or diagenetic, to distinguish them from the primary, or depositional ones. Secondary structures are discussed in this book, but selectively; only a few examples are shown. The reason is twofold: first, secondary structures are much more specific of a certain material, locality, or geologic unit than primary structures, and so are not easily amenable to categories. Only a few general types can be recognized. Second, they tell us less about the original environment of deposition, whereas primary structures also play the *primary role* in this respect.

As said before, both bedding planes (that delimit individual beds) and parting planes (that subdivide beds into constituent parts or laminae) are *physical discontinuities.* How, then, does one tell the one sort from the other? Is the distinction arbitrary (bedding surfaces separate beds by definition) or real? How is it applicable in outcrops and cores? To answer these questions, the concept of hierarchy must be introduced. Discontinuities in sedimen-

tation can be arranged in order of importance according to the duration of interruption in deposition, the amount of erosion or dissolution that occurred along them, the evenness of the discontinuity surface, and other criteria.

In principle, bedding surfaces mark longer interruptions than lamination surfaces do, and are thus of higher rank; this should be reflected in their being more sharply defined in the field. Unfortunately, such is not always the case. A bedding surface separating two beds of the same lithology can be more subdued than the surface separating two laminae of slightly different material. It thus seems that some subjectivity inevitably colors any attempt to recognize a hierarchical order among objects that involves some degree of abstraction. If you are on a trip with a geological party, do not wait too long to test this statement: ask everybody, in front of a suitable outcrop, to make a distinction between groups of beds, individual beds, and bed portions, and to sketch them in notebooks. Repeat the test at several outcrops, and note how frequent are the disagreements between "observations" (or, better, observers).

In spite of this limitation, let us expand a little more on the matter of bedding hierarchy. The bed (layer, stratum) is our essential reference. It represents the base line for both *stratigraphy* (where it is the building stone of stratigraphic successions or sequences) and *sedimentology* (it is a container of sedimentological features, including structures, and a framework for their observation). In classical stratigraphy, the bed has two characterizations: one is geometrical, the other lithological. In terms of geometry, the bed is a three-dimensional object in which two dimensions prevail on the third one (thickness); on the other hand, it is formed by a definite rock type, which implies that changes of lithology occur along bedding planes. Though a geologist's eye may be struck, in different circumstances, more by one or the other of the two aspects (the container or the content), lithology is commonly privileged in descriptions. In a succession where, for instance, sandstones alternate with shales (figure 8A and B) or with conglomerates (figure 9), geologists speak of sandstone beds and shale (or conglomerate) beds.

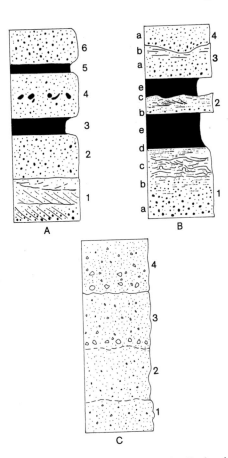

Sedimentologists, however, have introduced a *genetic* concept of bed and bedding based on interpretation of observable features like geometry, lithology, textures, structures, fabrics, and fossil content. In synthesis, these features constitute the *facies* of a bed or a group of beds. Facies is also a classic, albeit disputed, concept of stratigraphy, and has been continually refined with the progress of knowledge about sedimentary processes in Modern environments. Today scientists know a lot about these processes, whereby the definition of beds can lie on them, i.e., on the mode of bed emplacement. In this respect, a bed becomes an episode, an *event* of sedimentation: it is finite both in space (areal extent and thickness) and time (deposition occurs between two pauses). Its limits may coincide with lithological changes but are not defined by them; it is the process of emplacement, not the materials emplaced, that defines the base and top of a bed. Referring to figure 8, suppose that the process is represented by turbidity currents; these flows transport suspended particles of different size, which are stirred and mixed by

Figure 8. Definition of stratification units (*bed* and *layer*) in columnar sections of clastic sediments; numbers refer to beds, lowercase letters to internal subdivisions (when present) characterized by different structures. **A** and **B,** alternating sandstone and shale beds. **C,** sequence of amalgamated pebbly sandstone beds. Bedding surfaces are barely visible in **C** and do not constitute physical discontinuities (e.g., a barrier to vertical migration of fluids) as in **A** and **B.**

Figure 9. A set of amalgamated beds in Ancient fluvial deposits (Devonian Old Red Sandstone, Midland Valley, Scotland). Sandstones and conglomerates alternate vertically and, in part, laterally. Beds are roughly parallel but lenticular and wedge-shaped in detail. The hierarchical relation between bed and laminae can be appreciated here. The central bed of conglomerate, for instance, is subdivided by diagonal lines into crude, thick laminae. The laminae form a single set extending to the whole bed. The underlying bed of conglomeratic sandstone is composed of several sets of cross-laminae cutting each other, i.e., a coset. The generic term cross-bedding, or cross-stratification, can be applied to it (to indicate the style of bedding but not its hierarchical organization). Photo G. G. Ori 1992.

turbulence. When the flow slows down, the particles settle in order of decreasing size and weight, forming a vertically sorted deposit, i.e., a *graded bed*. Sand is concentrated at the base, mud at the top; the lower part of the bed is sandy, the upper one is muddy. Diagenesis transforms them into sandstone and shale or mudstone, respectively.

The alternative view is to consider two distinct beds, one made of sand and the other of mud. Which is the best view? By applying the lithologic criterion, you have two beds for each sedimentary event; with the sedimentological method, there is only one bed, whose lithology changes gradually from base to top.

One can wonder whether this is a relevant problem, or a pure question of nomenclature. The important point here, even for practical purposes, is to know whether a sedimentary body was made by one or more events. If we know what the case is, we can more easily make predictions about its lateral changes of geometry and lithology, and give a more correct interpretation of its environment of deposition.

Assume, for example, that all mud drapes over sand beds were eroded prior to deposition of every next sand; what you eventually get is a monolithic body of sandstone made of similar, welded beds. If you trace it laterally, you can see shaly intercalations between sandstone beds appear and grow thicker and thicker; this is because you go from the place where all mud was eroded to that where the same mud was deposited. By comparing directly the successions at the two ends in lithological terms only, one would say that the sandy sequence is made of just one sandstone bed, the other of many sandstone beds. However, the number of depositional events is the same: only their mode of preservation changes from place to place.

The welding of beds of the same lithology requires that one looks at details and uses all criteria of facies differentiation to recognize bedding: every new bed can be marked, for example, by a change in grain size, or porosity, or cementation. This is not necessarily so, however: different beds can be made of the same materials, and be virtually indistinguishable. The phrase more commonly used for a pile of monolithic beds, usually separated by erosional surfaces, is *amalgamated beds*.

To avoid confusion in describing and commenting on the illlustrations in this atlas, the following convention was adopted: when I speak of individual objects, *bed* is used in a lithologic or otherwise generic meaning, *layer* in the sedimentological meaning. The collective terms *bedding, stratification,* and *layering* are used interchangeably. A layer often corresponds to a lithological couple or, less commonly, a triplet; a classical turbidite, or a tidal layer (a deposit made by the waxing and waning of a tidal current), is thus composed of a sandy bed and a shaly bed.

A hierarchical classification of bedding has been set up by Campbell (1967); it has stood the test of time and is followed here. A *bedset* is the local, vertical expression of a sedimentary body made of several beds, or layers; when not otherwise specified, a sedimentary body is meant as a multilayered unit, i.e., a bundle or sequence of depositional events. If, observing the set from the bottom up, you notice a certain order, a kind of rule in the succession of strata, the set can be called a *sequence* or a sedimentary *cycle* (there is no well-defined and agreed-upon difference between these terms). A sequence can be defined by one or more parameters: if the thickness of individual beds is used, the sequence can be qualified as *thinning-up* or *thickening-up* (figure 10); if the grain size is preferred, analogous terms are *fining-up* (figure 11) and *coarsening-up* (figure 12). The combination of two bedsets with opposite trends gives a symmetrical sequence ($-/+$ or $+/-$; see figure 10), with an obvious cyclical pattern. On a descriptive basis, the asymmetrical bedsets could be termed sequences (upper row of figure 10), the symmetrical ones termed cycles.

What you observe, however, is not necessarily a continuous record of sedimentation, but what is preserved of it; a complete cycle could have been deposited, then erosion took away the upper half. In this way, an asymmetrical sequence can actually record a truly cyclical phenomenon. This is probably the main reason why the difference between sequence and cycle has never been convincingly

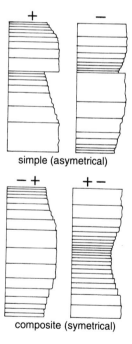

simple (asymetrical)

composite (symetrical)

Figure 10. Bedding sequences and facies sequences (facies are represented by types of beds) are bedsets with a certain vertical order within them. This vertical trend is shown by a single parameter (bed thickness, grain size) or more. Facies characters can be synthesized by particular *indexes* (e.g., ratios between parameters: sand to mud, clastic versus nonclastic components, etc.), which can also be useful in revealing trends. Here, the thinning-up trend is symbolized by a plus ($+$) ("positive" sequence), the thickening-up trend by a minus ($-$) ("negative" sequence).

Figure 11. A bedding sequence defined by grain size (fining-up, = FU) in alluvial sediments: it starts from coarse gravel, ends with silty sand ("dirty sand") and is capped by a soil horizon. Beds are amalgamated and almost undistinguishable in the lower part. Photo taken in quarried deposits of the river Reno near Bologna. PHOTO G. G. ORI 1992.

Figure 12. An example of thickening and coarsening-up ("negative") sequence. The story is complicated here by the occurrence of nonclastic deposits at the top (evaporitic gypsum in large crystals). Gypsum is present also in the middle but is detrital; thin beds at the base of the sequence are siliciclastic (siltstones and fine sandstones). Dome-shaped deformational structures are present at the top. This sequence is one of seven horizons of "Upper Gypsum" in the Messinian of western Sicily.

codified, regulated, and accepted by the geological community (in spite of many attempts to do so, "sequence stratigraphy" being the last and most comprehensive "legal" effort). Some ambiguity also exists because these terms are employed for both the real thing (the beds, the rocks) and the abstract concept (implying the controlling process or the geologic age, i.e., the time span during which the beds were deposited: eustatic cycles, tectonic cycles, lower Cretaceous cycles, etc.). Perhaps, it is more precise to use them as adjectives: *sequential* bedsets, sequential arrangement of beds, cyclical bedsets, and so on. *Asequential* bodies, by contrast, are those devoid of an evident vertical order in bedding; in them, changes of bed thickness, lithology or grain size appear to be random, at least in the field (subsequent processing of data by statistical methods can reveal a hidden cyclicity).

Sedimentary beds, layers, and strata cannot be described in a purely qualitative way—a specification of thickness, for instance, is required. This can be done by direct measurement or by referring to classes of thickness (see table 1); a scale, in centimeters or meters, must be chosen.

Parting surfaces within beds, although subordinate in importance to bedding planes, are sometimes sharp-cut and look like them; the distinction is not always easy, especially for beginners; some practice is needed to get a certain familiarity with these phenomena. Experience tells that the more you learn about sedimentary structures, the less difficult it is to understand bedding. A bed can be completely structureless or structured; in the latter case, structures appear in the form of *laminae* of variable geometry. Lamination can occur throughout the whole bed or be localized in *laminated intervals;* one or more of these intervals, or divisions, can exist within a bed. A single laminated interval often appears in the upper portion of a sandstone bed (figure 6), which is a useful way-up criterion where the strata are strongly inclined or subvertical.

In between the individual lamina and the laminated interval, there is an intermediate rank, occupied by the *laminaset;* this is a genetic package of laminae that is, anyway, distinguishable from adjacent ones on a reasonably objective basis (for example, by scour surfaces). The

Table 1.

Thickness scale for beds (layers) and laminae (in cm)

Beds and Layers	Laminae
>300: extremely thick[a]	3–10: very thick[b]
100–300: very thick[c]	1–3: very thick
30–100: thick	0.3–1: thick
10–30: medium	0.1–0.3: medium
3–10: thin	<0.1: thin
<3: very thin[d]	

[a]Avoid the prefix *mega* if you do not know the overall volume of beds; to estimate the volume, you must have an idea about the areal extent of the bed and its thickness changes. *Do not judge the size of a bed from one outcrop!*

[b]Use the terms "faintly laminated," "diffuse lamination," "banding."

[c]In a certain usage, beds thicker than 1 meter are qualified as *massive;* this term is not recommended as it is also used to indicate beds devoid of internal structures.

[d]Many authors, following McKee and Weir (1953), put a lower limit of 1 cm to beds; units thinner than 1 cm are called laminae. Here, instead, a genetic-hierarchical criterion is adopted for distinction (a lamina is part of a bed, i.e., a subevent or elementary event.)

genetic link among laminae in a set is demonstrated chiefly by their parallelism, or *conformity,* which records an episode of continuous sedimentation. A single set of laminae can form the whole laminated interval (see figure 8.B2); otherwise, it consists of bundles of sets (*cosets:* figure 8.A1; figure 9, lower half).

Are there lower and upper size limits to sedimentary structures? The matter is not yet settled, and at the moment precise limits are not established. Ordinary structures are macroscopic characters, which means that they are recognizable with the naked eye. Morphological features that can be observed with the help of a hand lens or a microscope (for example, on the surface or interior of sedimentary particles: see figure 13) are more properly called microstructures. Only a passing mention is here made of them as they are the object of other kinds of atlases (thin sections of rocks, SEM images).

Half-way between micro- and macrostructures one can consider the laminae, thin sheets of sediment that are often less than 1 mm thick (paper-thin). Laminae are the smallest sedimentation units; they represent an almost continuous deposition, interrupted only for some instants or just varying in rate.

As for the upper end of the size spectrum, no stakes have been posted. To get an idea of the largest objects to be found, imagine parachuting to a sandy desert (an *erg:* see plate 24). From the plane, before the launch, you see a rippled surface, quite similar to that you tread upon in a beach or tidal flat. The scale, however, is different. The relief, for example, is in the order of meters or tens of meters instead of centimeters, as you can see from the ground. This wavy topography, extended over wide areas, reflects the shapes of eolian dunes. Dunes are made of well-sorted fine sand (about the size classification of

Figure 13. Magnetic spherules (micrometeorites) in a SEM (scanning electron microscope) image. Microstructures, related to melting, corrosion and rupture, are visible on their surface. The spherules were extracted from a pelagic rock of Mesozoic age. Meteorites of very small size survive the encounter with the terrestrial atmosphere better than larger ones do. They continually fall all over Earth but are widely dispersed and deteriorate. A concentration of micrometeorites can occur in pelagic sediments because of their slow rate of deposition. From Castellarin and others 1974; see plate 171.

sedimentary particles, see table 2). The sand is transported and accumulated by the wind, whose power and persistence determine the dune size. On the upwind or stoss side of a dune, the sand grains are in transit, because a high shear stress is applied there and favors erosion or nondeposition; deposition occurs on the downwind or lee

Table 2.

Grain size of sedimentary particles: the Udden-Wenworth scale

Size classes (mm)	Components	Aggregate
>256	cobbles and boulders	rubble
256–4	pebbles	grave
14–2	granules	—
2–1/16 (0.063)	sand grains	sand
subdivisions		
2–1	sand grains	very coarse sand
1–1/2	sand grains	coarse sand
1/2–1/4	sand grains	medium sand
1/4–1/8	sand grains	very fine sand
1/16–1/256 (0.004)	silt grains	silt
<1/256	clay particles (minerals)	clay

side, the steeper one. Overall, the dune surface is a depositional surface as the net result of the wind action is deposition, not erosion.

Generally speaking, the morphology of sediment-fluid interfaces is ascribed to one or more types of sedimentary structures, in particular to so-called *bed forms*. In fluid dynamics language, the topography of a depositional surface is called roughness: it influences the behavior of a fluid moving over it and, at the same time, is influenced by the flow. But dunes are also a form of terrestrial relief, i.e., geomorphic features. Dunes are thus studied by sedimentologists, hydraulic engineers, and geomorphologists. Moreover, dunes have a considerable mass and thus can be regarded as sedimentary bodies: if you dig a trench in one of them, you will see several superposed beds. Sedimentary bodies are mappable units and are studied by stratigraphers.

In conclusion, what is a dune: a sedimentary structure? a geomorphic unit? a stratigraphic unit? It can actually fall into all these categories: it depends on the approach one chooses. The tendency to pigeonhole natural objects is in our brain and education, not in nature. In the academic world, disciplines and specialization proliferate and, consequently, it is normal that the same objects are contended between different specialists. No doubt, anyway, that dunes are sedimentary structures, among other things: they are formed by the same processes that produce the much smaller ripples. From the genetic point of view, there is only a difference of scale.

Problems arise, as usual, when one tries to define and classify things. In this case, you can see that, at the smaller end of the scale, structures are accessories, ornaments of beds; at the larger end, subaerial dunes, and even greater structures discovered under the sea, have the same extent of beds, and can bound many of them. For convenience, I shall speak of large-scale structures when I focus on the shape, the morphology of large sedimentary objects; I shall treat them as bodies or *stratigraphic units* when my focus is on their content, *architecture* (internal organization), and volume.

Some large-scale structures are erosional, and truncate previous deposits; such a truncation is shown by stratigraphic sections as an angular or irregular contact cutting through one or more beds. An erosional surface of large scale is a *stratigraphic discontinuity* whose rank depends on what happens above it: if it is covered by just one bed, i.e., a single sedimentary event, it is equivalent to a bedding surface. If, on the other hand, several beds abut against it, the surface has a higher rank, and must be regarded as an *unconformity:* it remains exposed for a significant geologic time before being buried by new strata. Unconformities have a regional extent: this means that a single erosional structure does not suffice to produce an unconformity. Several of them must be juxtaposed to form the complex topography that characterizes such regional surfaces; they are, in essence, subaerial or subaqueous landscapes of the past.

Large-scale structures and sedimentary bodies are rarely visible in their entirety in stratigraphic sections; the size of most outcrop is insufficient to show them. They are best displayed by seismic sections of the subsurface, although one must be aware that a seismic image has its own drawbacks: it is the result of instrumental processing, filtering, etc., of elastic waves passing through receivers, and is more or less distorted due to the use of different scales for height and distance ("vertical exaggeration").

When one looks at sediments from a short distance, one's attention is attracted not only by structures but also by textures, i.e., by various characters of the sediment components such as coarser and finer particles, their shape, orientation, and spatial arrangement. The main textural parameter, especially in *clastic,* or *detrital* sediments (those that are carried and deposited by mechanical forces) is the grain size: it can be inspected visually, by comparison with grains mounted on slides or printed silhouettes, to determine the size class of the most abundant (modal) grains. An appreciation of the size range, or sorting, can also be cursorily made. If one wants to be more precise, one will take samples to the laboratory and do grain size analysis with various methods.

Notice that the terms coarse, medium, and fine are not used loosely; they match precise size ranges according to the reference scale (table 2). This scale is based on powers of 2 and is employed also for materials erupted explosively by volcanoes (*pyroclastic* deposits), with some modifications for the specific setting (table 3). Pyroclastic materials, although volcanic in origin, are emplaced by gravity and superficial processes like normal sediments; beautiful structures can be found in them, of which several examples are illustrated in this book.

Sedimentary structures, although preserved in sediments, do not record only sedimentation events or depositional mechanisms. Even in sedimentary environments, where deposition is predominant, erosion can occur, sporadically or with a relative frequency. Erosional structures, consisting of surfaces of variable shape and size, are thus produced; eventually, they are mantled by sedi-

Table 3.

Grain-size classes of pyroclastic deposits (from Schmid 1981, in Fisher and Schminke 1984)

Size (mm)	Components	Aggregate
	pyroclastic particles (general)	*tephra, pyroclastic deposit (rock), pyroclastite, tuff, tuffite* (general)
>64	blocks, bombs, cinder	agglomerate, pyroclastic breccia
64–21	lapilli	lapilli tephra, lapillistone, lapilli tuff, lapilli tuffite
2–1/16	grains, shards	ash, ash layer, tuff, tuffite
<1/16	shards	fine ash, fine tuff

ment and fossilize. It is very important to recognize erosional surfaces, especially of medium to large scale, because they represent breaks in sedimentation and loss of stratigraphic record. These surfaces usually appear as lines in cross-sections; at places, they are also exposed at the top of *substratal* beds or, as molds, at the base of covering or *filling beds*. Remember that, for a correct description of rocks, surfaces must be discriminated from volumes; strata filling a fossil channel, or paleochannel, should be quoted as a *channel fill*, not simply as a channel, as is often incorrectly done.

In a stratigraphic section, geologists should speak of a channel, or valley, only with reference to *erosional* surfaces which should be: 1) visible on the outcrop and not just inferred, and 2) correctly interpreted (not all erosional features are channels). To recognize erosional phenomena in stratigraphic successions, the application of the actualistic principle is essential: one needs to observe Modern landscapes, the action of the various agents that model them (streams, wind, glaciers, waves), and the results of this modeling in terms of topographic and geomorphic forms.

Besides depositional and erosional structures, deformational structures are found in sediments. The deformations I am talking about are those occurring in the early stages of the burial history, when sediments have undergone little diagenetic changes and are still soft (they may be cohesionless, i.e., made of loose particles, or coherent when particles stick together). To distinguish this kind of deformation from that produced later, by tectonic stresses, scientists use the term *soft-sediment deformation,* or *penecontemporaneous* deformation. It can be caused, for example, by creeping or sliding on slopes, expulsion of water, liquefaction induced by seismic shocks, shrinkage due to dehydration, etc. Practical criteria for discriminating syn-sedimentary from tectonic deformation are difficult to generalize and are not discussed here; they will be recalled case by case in a section of the atlas. I only remark that deformational structures have the same variability of scale as other types of structure.

The last group of sedimentary structures includes those produced by the activity of organisms: they are called biogenic. Organisms play two contrasting roles concerning sediments: constructive and destructive (or deformative). On the one hand, they contribute to the accumulation of sediments with their remains, both hard (shells, skeletons) and soft (tissues, cells, organic matter). On the other, plants and animals use soil or sediment as their home or shelter, as a source of food or simply as a route or resting place: in some way or other, they disturb the sediment and often obliterate structures previously formed in them. This kind of sediment disturbance is generally called *bioturbation* (or bioerosion if it affects hard rock or lithified sediment). Organisms do not leave their remains but the traces of their activities; when they are identifiable, these marks are called *trace fossils;* when they are not (being muddled in a wholly mixed sediment), the generic term bioturbation, or bioturbated bed (deposit, facies), is used.

The constructive role of organisms can be *active* or *passive* in relation to accumulation of sediment and growth of sedimentary bodies. Coral and algal reefs are examples of active construction (bioconstruction); accumulations of shells and shell fragments (bioclasts), produced for example by storm waves and currents, represent passive growth. In any case, depositional structures can be produced, such as accretionary laminae (i.e., stromatolites, discussed later).

We have thus reviewed the principal groups of sedimentary structures, which correspond to the main divisions, or sections, of the book. Specific mechanisms and modes of formation of the structures will be discussed in the explanatory text accompanying the plates. It seems expedient, however, before concluding these introductory remarks, to give some general information about sedimentary processes and environments, which are responsible for the origin of sediments and their structures. Mention will also be made of the use of structures in "facies analysis," as the type of stratigraphic analysis aimed at reconstructing paleoenvironments of deposition and erosion is termed.

Among sedimentary processes, the most important are those called *physical,* or mechanical. They are almost ubiquitous over the Earth's surface, and move huge amounts of solid matter from erosional to depositional domains. Moreover, physical processes can entrain and remobilize sediments accumulated by other processes; every natural or artificial material can thus become a component of clastic sediments. It is, therefore, logical to presume that a large amount of structures preserved in sediments are physical in origin.

Chemical and biological processes must not be overlooked, anyway. Not only do they play an essential part in controlling the Earth's climate, the ecosystems and the interactions between biosphere, atmosphere, idrosphere, and geosphere (through biogeochemical cycles, for example, or the removal of carbon from the atmosphere), but also build up sediments. In the geologic past, accumulation of dead organic matter and biochemical reactions have formed imposing coal beds and oil-fields, and even larger masses of calcium carbonate have been extracted by organisms from the water of seas and lakes. Carbonate rocks may show specific structures beside the whole range of types usually found in clastic rocks. After being segregated from water and precipitated in the skeleton of organisms, calcium carbonate is subject like any other material to physical disgregation, transport and redeposition: it becomes a clastic carbonate sediment. The same can be said for evaporites, the main representative of chemical sediments: they form, in the first place, by precipitation of salts from sea or lake water, because of evaporation. After that, evaporitic salt crystals can be removed from their original place and resedimented as clastic particles.

The finding of physical structures in evaporite beds gives us evidence that such a reworking indeed occurred.

Physical processes of transport and sedimentation are to be considered from several points of view. A first is that of the operator, or the *transporting agent,* such as wind, various types of aqueous currents, waves, and gravity flows. A transporting medium is not always needed; when gravity directly acts on solid particles, they fall, slide, or roll down slopes, and accumulate at their base. A planet devoid of atmosphere or with a thin one, like Mars, can have sediments of this type. On Earth, the atmosphere and the water masses can either act as *working fluids* that entrain and carry sediment to its resting place or as passive "spectators" of what gravity is staging (this means that part of the fluid is entrained *along with* the solid particles moved by gravity, while the rest of it stands still). In some way or the other, the energy that is dissipated by physical processes is gravitational, but a distinction is made between processes directly promoted by gravity (*gravity-driven*) and all others, in which a fluid acts as a go-between (*fluid-driven*). So-called sediment gravity flows belong to the first category.

You are now prepared to examine, more concretely, the ways sediments are removed, transported, and deposited: fast or slowly, abruptly or gently, a few particles at a time or en masse, etc.

In some cases, sedimentary particles travel at different velocities, or some are moving while others remain stationary: this occurs as well in a fluvial current, in a wind storm or under the waves approaching a beach. The main differentiation occurs between *suspended load* and *bed load,* i.e., between particles that have lost contact with the ground and are supported by fluid eddies, and particles that struggle along with difficulty near the bottom, pulled and pushed by the flow. The former are lighter, smaller and faster (they move at the same speed as the fluid, and make it turbid), the latter are heavier, coarser and much slower (also because of frequent collisions and strong friction). Individual particles in the bed load do not move continuously but come to rest temporarily; their movement can be described as a stop-and-go. They are also subject to a more intense wear (fragmentation due to collisions, abrasion). The moving grains can either roll, slide or jump; all these mechanisms (the last one is called saltation) are encompassed by the term *traction,* or tractive process.

Particles behaving differently in the same flow constitute distinct sediment populations and will be deposited in different places; the flow selects these populations from a parent stock produced by weathering in source areas. Not all processes have the same efficiency in doing this job: thereby scientists distinguish, among them, *selective* from more *massive* types. Sedimentary structures produced by selective flows (which can include waves) are called *tractive structures* if originated in the bed load, *fall-out* or settling-related structures if generated by a passing or stationary suspension. The former are usually found in coarse sediments (gravel to sand size), the latter in fine

materials (silt to clay). Traction and fallout can also be combined during deposition of particles from highly turbulent suspension flows; the grain sizes involved are intermediate between the coarser and finer end members (fine sand to coarse silt). Traction-plus-fallout structures are common in turbidites and deposits of fluvial floods and melt water.

Mass transport processes put in motion in a short time, often instantaneously, consistent volumes of previously deposited sediment or detritus covering weathered rocks. This material can be quite heterogeneous, from stones to mud, to plant stems or branches and other debris, or more sorted (only mud or sand, for example). It can mix with water and form a viscous slurry, or move like a dry avalanche. Dry or almost so mass flows occur on the flanks of volcanoes, when ejected particles mix with hot gases to form dense clouds, heavier than the surrounding air and driven downslope by gravity. If water is involved in the eruption, it vaporizes.

Most mass flows are gravity-driven and occur on relatively steep slopes, both subaerial and subaqueous. Their speed is related, on one hand to the topographic gradient, or steepness of the slope, and to the mass involved; on the other, to the concentration of solid particles (the ratio between their volume and the volume of fluid), their specific weight (which regulates their buoyancy) and their relative, or specific surface (ratio between surface and volume, which increases with decreasing grain size and affects frictional resistance).

Flows in which solid particles are relatively diluted in the fluid mass move more easily, can develop turbulence and segregate their load into subpopulations of grains. Consequently, the ensuing deposit can be structured, at least in part: vertical size grading and laminae of different styles can be found in it (which led to the definition of the "Bouma sequence," figure 8.B). For higher concentrations of sediment, the movement is slowed down by the stronger friction, both within the mass and at its boundaries; the behavior tends to be that of a visco-plastic material, a sort of toothpaste, and is common to water-soaked rock debris, glacier ice and flows of viscous lava. The finer particles mix intimately with the fluid to form a cohesive phase (like mud in water) capable of supporting, with its strength, coarse and heavy particles: large blocks can thus be transported at the same speed as the smaller clasts and the mud (or equivalent matrix material). As all these particles are entrained all together, helter-skelter, they are also deposited in the same way. Nor is it matter of real deposition: the flowing mass reduces its speed and "freezes"; freezing is equivalent to the solidification of a lava, and is caused by a decrease in slope gradient, water expulsion, or both factors (the reason is that motive force wanes and internal friction increases).

Less dense and turbulent mass flows look like fast moving, turbid clouds: on land, they take the aspect of dry avalanches on mountain slopes, or of "nuées ardentes," pyroclastic flows and base surges on the sides of volcanic cones. Sand storms in deserts and major river floods (especially the "flash floods" of torrential streams) can have this character but are generally slower. Under water, these phenomena have been inferred or detected instrumentally but never observed, except on a small scale: sediment is remobilized in shallow water by strong storm waves or tsunami waves, and redeposited there as *storm layers* or entrained in deep water by a gravity-driven suspension cloud, a *turbidity current*. The excavation of deep-water channels and canyons, the buildup of sediment in their levees and the mantling of vast areas of the abyssal seafloor by sand and mud containing coastal and fluvial debris, are all effects of the action of these density currents, whose deposits are called *turbidites*.

The slower and more concentrated mass flows (sometimes called hyperconcentrated) can also occur both on land and under water: in certain cases, they are monogenic, i.e., made of a single component (mud, sand, stones, wet snow: the coarser the particles, the stronger is the solid friction, and the steeper the slope needed for movement), but more commonly they are constituted by a mixture of two or more textural types (*modes*). In the latter case, the term *debris flow* is used, with the exception of pyroclastic deposits: hot ash directly emanating from volcanic vents is emplaced by *pyroclastic flows* (denser, highly concentrated variety), whereas cooler ash remobilized from slopes by rainstorms or simply gravity form so-called *lahars*.

Debris flows occur in both subaerial and subaqueous environments; in the first case, they can be directly observed, for example, on mountain slopes and streams after heavy rains and thunderstorms. Cohesive flows move like a paste, while a higher liquid content fluidizes the mass, which nonetheless keeps a high viscosity and a laminar behavior.

Mass flows pick up their material from a *repository* localized in a source area of some kind (snow on mountains, talus or scree debris on valley sides, river sediment on deltas). It is thus implicit that some other sedimentary processes (or gravity alone) accumulated this detritus in the first place. This is a first requisite for mass flows to occur: a causal factor acting in the long term. Another requisite is the potential instability of the accumulated material: the pull of gravity on slopes, the lubricating effect of melted snow, the excess pressure of water in pores, the delayed compaction due to rapid deposition are examples of factors that contribute to the instability of sediment. The last requisite is the "final push": a sudden overload, a seismic shock, a heavy storm, etc. This *immediate cause,* or "trigger," is not always necessary: the least tremor in the ground or perturbation in air or water can induce a catastrophic flow only because the other factors had prepared suitable conditions for it. The return or *recurrence time* of mass flows, in fact, has more to do with the size of the available reservoir and the energy accumulated in it (and hence the time required for its formation) than with the frequency of triggering events.

Mass flows fall in the category of *catastrophic* events. Some object to this term because of its affirmed anthropic

implications (it alludes to damages inflicted on human communities) and prefer "episodic," "sporadic," "exceptional," or "rare." In addition, there is still, among geologists, a prejudice against catastrophism dating back to the old quarrel with actualism. To be a catastrophist implied a philosophical view by which a divine power bears the responsibility for major natural changes. Actualists fought this theory, among other reasons, because it would induce an attitude of passivity and resignation toward nature. Today, however, there is no reason to attach symbolic or religious meanings to natural catastrophes: they do not reflect God's wrath, but natural causes, and a better knowledge of these causes would certainly contribute to alleviate their impact on society. Structures indicative of catastrophic events (e.g., earthquakes) recorded in sediments are stressed in my text, with a hint to their potential utilization for natural hazard studies and risk analysis.

Tractive currents, fair weather waves, sluggish suspension flows, and wind are examples of "normal," i.e., typical "actualistic" processes; the marks left by them in sediments consist in more structures, more order, more sorting of materials, and less erosion with respect to products of mass flows. Normal processes operate almost continuously, though varying in intensity with regular (tides, seasonal changes) or irregular periodicity. The energy at stake can be much greater than that developed by catastrophic processes (think of the enormous amount of energy stored in seawater and dissipated by tides every day), but it is less concentrated in both time and space. It is thus less available (which means less valuable, in terms of *quality*) to do mechanical work. That is the reason it is said that "normal" processes produce structures of lower energy than mass flows; this statement is not true in absolute terms, but relative to the energy spent on a particular sediment in a particular place.

The causal relationship between a sedimentary structure and a sedimentary process is not always simple or straightforward. This must be borne in mind to avoid mistakes and hurried conclusions. Some points are here suggested as *caveats:*

• a structure can be produced by different processes: a typical case is the intrastratal deformation known as convolute laminae (see plates 116–118);
• on the other hand, the same process, e.g., a subaqueous slide, can have different effects in materials with diverse physical properties and consolidation state;
• more than one process can combine to give a certain structure. They can act simultaneously or in a sequence, continuously or discontinuously;
• there are thus single event (monophasic) and multi-event (poliphasic) structures.

If the link between structure and process is complex and needs some thinking, even less definite is the link between structure and environment. Processes are just one of the factors and variables at play in sedimentary environments. In many of them, for example, fluids exert a tractive drag on sediment and form tractive structures;

consequently, tractive structures, per se, are trivial (however, their presence rules out massive and chaotic processes). There are, fortunately, certain structures that are more characteristic and significant than others, and will be stressed as *environmental indicators*. This will be done on a case by case basis, assuming that the reader has already a summary knowledge of what is a river bed, a beach, or a glacier, i.e., a natural environment in the common geographical sense.

Beside structures, there are other features in sediments and sedimentary rocks that can be utilized as indicators: composition, texture, fabric, etc. All these elements, which are comprised in the concept of facies, are valuable tools for the reconstruction of Ancient environments and geography (paleoenvironmental and paleogeographical analysis). A geologist cannot reproduce a landscape that has vanished but can infer and reconstruct it piece by piece, with more or less confidence, by using fossil evidence as a clue. The more indicators that can be discovered, the more reliable the reconstruction is.

Information on paleoenvironments and paleogeography allows us to better understand the history of our planet and the work of its systems: geodynamics and internal processes (magnetism, magmatism, seismicity, plate movements), external processes and influences (solar radiation, atmospheric and oceanic heat machines, climate, weathering, erosion, sedimentation), biological processes, and the ecosphere. This is a noble goal in itself, but not the only one for research.

Sedimentology can be applied to find resources (minerals, water, energy). The analysis of sedimentary facies, for example, can elucidate the characteristics, properties, and history of sedimentary basins, where diverse types of economic materials can be found.

One cannot trace the subsidence history of a basin without knowing the depth below sea level at which some critical stratigraphic levels were deposited, and this depth must be inferred through appropriate indicators. It is part of the "search for the paleoenvironment" game. Basin analysis also requires that the sources, pathways, and sinks of sediments be located. A major contribution to trace the dispersal of sediment in Ancient basins can be given by the sedimentary structures and the geometry of sedimentary bodies (in conjunction with compositional data that testify to the provenance of clastic sediments). The *orientation* of structures and deposits must be measured in this respect, both in relation to present-day spatial coordinates and to the trend of Ancient shorelines, margins of continents, mountain chains, volcanic ridges, and so on.

The geometrical attributes of sedimentary structures (shape, asymmetry, elongation, etc.) can be used to get the directions of paleocurrents, paleowaves, and paleoslopes. The orientation of structures is reported in maps and diagrams and is a current and essential tool in the sedimentologist kit for facies analysis. Since the classic book by Potter and Pettijohn (1963), the operational procedure is known, comprehensively, as *paleocurrent analysis*.

Let me conclude this introductory section with a remark that I deem necessary. The present book has been conceived to help you in identifying sedimentary structures, and to do so with a critical attitude. It is a guide to observation, not a catalog of identical, mass-produced objects. In geology, a recognition procedure is rarely easy and immediate because geological objects are complex, almost unique entities. They are very different from the more elementary objects of chemistry and physics: you cannot tell the difference between two atoms or molecules of a certain chemical species, but no two sand grains or two structures are ever exactly alike in all minimum details. Each of them has a certain degree of individuality, like persons.

For this reason, you must stop in front of each structure and think about it for a while. Do not be satisfied with the first explanation that comes to your mind, especially if it seems obvious. Make more than one hypothesis, weigh and balance them, use logic in discriminating between them. Make comparisons and use analogies (but do not rely only on them). Make sure that your interpretations are consistent with the facts by looking for more facts (even though they are never sufficient to definitely prove a point). Free your fantasy and do not be afraid to speculate, but check for internal consistency and coherence of your speculations. Several plates in this atlas are presented as examples to stimulate such a way of reasoning. Good reading, and good work!

References

Allen, J. R. L. 1968. On the character and classification of bed forms. *Geologie en Mijnbouw* 47: 173–85.

Allen, J. R. L. 1968. *Current Ripples*. Amsterdam: North Holland.

Allen, J. R. L. 1982. *Sedimentary Structures: Their Character and Physical Basis*. 2 vols. Amsterdam: Elsevier.

Bosellini, A., E. Mutti, and F. Ricci Lucchi. 1991. *Rocce e successioni sedimentarie*. Turin: UTET.

Bouma, A. H. 1962. *Sedimentology of Some Flysch Deposits*. Amsterdam: Elsevier.

Collinson, J. D. and D. B. Thompson. 1982. *Sedimentary Structures*. London: Allen & Unwin.

Campbell, C. V. 1967. Lamina, laminaset, bed, and bedset. *Sedimentology* 8: 7–26.

Cas, R. A. F. and J. V. Wright. 1987. *Volcanic Successions: Modern and Ancient*. London: Allen and Unwin.

Conybeare, C. E. B. and K. A. W. Crook. 1968. *Manual of Sedimentary of Sedimentary Structures*. Australian Dept. Nat. Dev. Bur. Mines Res., *Geol. & Geoph. Bull.* 102.

Fisher, R. V. and H.-U. Schminke. 1984. *Pyroclastic Rocks*. New York: Springer-Verlag.

McKee, E. D. and G. W. Weir. 1953. Terminology for stratification and cross-stratification in sedimentary rocks. *Bull. Geol. Soc. Amer.* 64: 381–90.

Pettijohn, F. J. and P. E. Potter. 1964. *Atlas and Glossary of Sedimentary Structures*. New York: Springer-Verlag.

Portmann, A. 1965. *Aufbruch der Lebenforschung*. Frankfurt am Main: Suhrkamp Verlag.

Potter, P. E. and F. J. Pettijohn. 1963. *Paleocurrents and Basin Analysis*. New York: Springer-Verlag.

Ricci Lucchi, F. 1970. *Sedimentografia: Atlante fotografico delle strutture primarie dei sedimenti*. 1st ed. Bologna: Zanichelli.

Shrock, R. R. 1948. *Sequence in Layered Rocks*. New York: McGraw-Hill.

Note. A more exhaustive list of references, based on specialized literature, was prepared but, for reasons of space, was not included here. It was published in the Italian journal *Giornale di Geologia* (vol. 52, 1991) and is available on request at the editor's address: Dipartimento di Scienze Geologiche (Department of Geological Sciences), Via Zamboni 67, 40127 Bologna, Italy. Every reference is accompanied by abbreviated information concerning the quality of illustrations, type of data on structures (outcrops, cores, seismic sections), sedimentary environment and facies of the unit where the structures come from, and possible experimental setting (flumes, etc.).

Some additional references will be found as endnotes in the plates section.

CHAPTER 1
Geometry of Bedding and Sedimentary Bodies

Plate 1
Plane-parallel bedding: panoramic view

Before reviewing the various types of sedimentary structures, beds and bedding are introduced as they show up at various scales, both on the Earth's surface (outcrops) and subsurface (instrumental records of acoustic waves, or seismic sections).

The term *scale* in rock outcrops means two things: the size of the exposed section, and the distance from which it is observed. The example shown here is the side of a mountain (Mt. Carpegna in the Apennines, Italy) and represents a large-scale outcrop, indeed one of the largest possible. Several hundred meters of superposed, parallel strata are visible in the picture, and many of them can be traced laterally for more than 1 km (we say they are *continuous* within the outcrop limits). The *style* of stratification is thus clearly expressed, whereas the resolution of the image is not sufficient to tell whether the lines you see, which represent the intersection of *bedding planes* with the topographic surface, mark *individual beds* or bundles of beds (*bed sets*): geologists can speak of *levels* or *horizons*. This problem of resolution is the same as you have in seismic sections (see plate 2), which means that the scale is comparable. Although even a very large outcrop covers only a part of the area of a subsurface section, it gives you an idea of what would appear if you could dig a trench with a gigantic shovel in the fill of a sedimentary basin.

The principles of original horizontality and of superposition, mentioned in the introduction, are well illustrated by this picture. The pile of strata grew regularly, one bed after the other, in the most classical way you can imagine sedimentation to occur. There are, anyway, other *growth models* for bedding and sedimentary bodies: this one is identified as *vertical accretion, aggradation,* or, more colloquially, *layer-cake.*

The formation to which this outcrop belongs is named after another locality of the Apennines (Monte Morello, near Florence); this means that this section is not the *type section,* which serves to define, according to stratigraphic rules, a lithostratigraphic unit such as a formation. In Alpine chains, geologists still use also names for categories of formations, e.g., *flysch* and *molasse.* These terms refer to assumed tectonic and sedimentary conditions of deposition in different development stages of basins associated with fold-thrust belts (*orogenic basins*). The M. Morello Formation is a "flysch" of Paleocene age, made mostly of carbonate turbidites that filled a subsiding abyssal plain and were later detached from their oceanic substratum by tectonic compressions that built the Apennines chain.

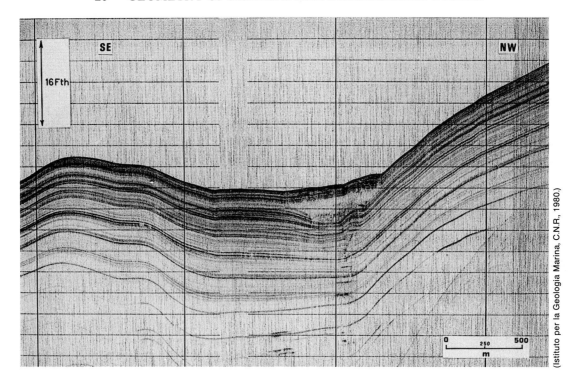

(Istituto per la Geologia Marina, C.N.R., 1980.)

Plate 2
Parallel beds: draping

This seismic section was recorded by a device called subbottom Profiler working with waves pulsed at a frequency of 3.5 kHz. It shows the subbottom (strata underlying the sea bottom) of the Gulf of Taranto in the Mediterranean Sea. The strata are here manifested as *reflectors* outlined by differential reflectivity of their surfaces; in other words, acoustic waves sent from above by a ship are reflected and more or less absorbed according to contrasts in physical properties (particularly density and elasticity) between beds of the sedimentary pile. All the reflectors shown here are depositional surfaces (there is no evidence of erosion).

The style of stratification is similar to that seen in plate 1 in what concerns the *parallelism* and lateral *continuity* of depositional surfaces. In contrast, bedding is not, for the most part, horizontal, as it covers sloping surfaces. Notice that the angle of slope is higher than the real one, due to the *vertical exaggeration* of the seismic image: you can see that the vertical scale (Fth = fathoms; 1 fathom = ca. 1.83 m) is markedly different from the horizontal one.

At the junction between horizontal reflectors at the center and inclined reflectors to the right, a "transparent" lens occurs, with undulations above and below it. If you follow the reflectors immediately underlying the lens from the flat to the sloping area, you can note that the sediment thickness decreases. This means that the lens is an extra accumulation of sediment derived from the adjacent slope, where it has left a gap in the record. The lens is not bedded because this sediment slid, broke and mixed with water. This lack of structure, or *disorganized* character, is thus interpreted as a *gravity deposit,* i.e., a submarine slide or a mass flow.

Aggradation is the predominant growth style, as in plate 1: when it occurs on a rough bottom and mantles it with a relatively uniform thickness of sediment, the more specific term is *draping.*

Photo: Institute for Marine Geology, C.N.R. 1980.

Plate 3
Parallel bedding: close-up view

Skipping several orders of magnitude in scale, we approach here bedding from a distance of a few decimeters (notice lens cap at the bottom of the picture). Fine geometrical details are thus observable, along with the lithology and texture of sediments. It can be seen that darker beds are more finely textured: they represent compacted mud (mudstone) alternating rather regularly with beds of fine sand and silt. Such an alternation of two kinds of sediments, or *lithotypes,* is called by some a *rhythmite.* The term alludes to the fact that there is a repetitive superposition of the same basic motif of deposition, a sort of rhythm. Each unit of this rhythm is a couple of beds (sand/mud), i.e., a layer. The thickness is in the order of centimeters (thin to medium bedding).

Two possible explanations come to mind for these couplets: 1) the mud settled, slowly and continuously, from a suspension when the water near the bottom was calm (if it is agitated, the particles are kept suspended). From time to time, a stronger current arrived, laid down its load of sand and silt, and decayed; 2) not each bed but each layer (couple) represents a waning current. Deposition started in every event with the heavier and coarser material, followed by the lighter and finer one. Nothing was deposited before the arrival of the next flow.

Under the first hypothesis, two distinct sedimentary events (a background of fine sediment rain punctuated by sporadic inputs of sand-carrying flows of higher energy) alternate in time and make the couplet; under the second assumption, only one event, or sediment inflow, makes a couplet, with a change in sediment texture occurring during the event.

Some sand beds have a wavy top, which reflects the profile of ripples. The implication is that, when a current capable of rippling the sand slows down and stops, the ripples are not canceled. The capping mud drape allows them to be preserved and to fossilize. Also, some deformational structures (see further on, plates 104 to 139) are visible at the base of a sand bed near the picture center: they consist of small lobes of sand sinking amid pointed crests of mud, and are classified as load structures.

Pleistocene lacustrine deposits near Todi, Umbria Region, Italy.

The basin where these sediments accumulated was a deep, subsiding lake formed by tectonic stretching (extension) of a newly uplifted chain (the Apennines). This kind of basin is generally called intra- or intermontane and characterizes late evolutionary stages of orogenic belts. This one was active in early Quaternary times (Pleistocene), and rapidly filled by sediments. The sediments were supplied by short-headed streams that drained nearby highlands; the influx was probably *seasonal*. No room was available to accommodate sedimentation along the lake border; therefore, the load of stream floods was emptied into a subaqueous slope where it formed density (turbidity) currents. Density flows could carry the sediment as far as the flat basin center. Seasonal rhythms of sedimentation, particularly in lakes, are also called *varves* after the Swedish term coined for glaciolacustrine deposits left by retreating glaciers after the last Ice Age in northern Europe.

Assuming that the Quaternary hydrological regime was similar to the Modern one in central Italy (two rainy seasons alternating with two drier seasons), every second couple would record a year. The two flood peaks might be, however, unequal, with one of them leaving no record. A single couplet should represent a year in this case. The conclusion is that one must be very cautious in defining (or, better, interpreting) varves and using sediments as geological clocks. It is safer to use the term *varvelike* for well and thinly bedded, rhythmic alternations deposited in calm water. Examples can be seen in plates 59 and 147.

Photo: G. Basilici 1992.

Plate 4
Bedding in sediment cores

Stratification is often ill-defined in sediments cored from the sea bottom. Only the superficial, still unconsolidated deposits are represented in cores. Moreover, corers do not usually penetrate thick sand layers. The difficulty in recognizing beds and layers in cores stems, then, from two reasons: the slight lithologic contrasts and the soft, prediagenetic state of materials (diagenesis causes hardening, and hardening is differential and enhances physical contrasts).

Cores are opened and photographed as a routine procedure in marine geoscience or, if X-ray equipment is available, they are radiographed before opening. The picture shown here is magnified as compared with that of figure 4 (see introductory section): the length of each segment is about 35 cm. The white borders are sawed plastic liners.

The sediment consists almost entirely of mud, except for some silty sand with laminations in the middle-upper part of the left segment. The sand is graded, and fades upward into mud; it lies upon a dark, sharp-based level rich in volcanic fragments. The overall graded unit is a turbidite layer, and its sharp base is probably erosional. Mud beds are distinguished by differences in color or tone, with darker hues indicating concentrations of organic matter.

Cores, of course, do not give us any information about the geometry of bedding; one can only appreciate vertical contacts and local thickness. In principle, a cored bed could be recognized and traced in a seismic section, but the bottom reflections tend to mask and "amalgamate" reflectors of the immediate subbottom, which are rarely resolved.

Coring site: deep sea plain of the Tyrrhenian Sea.

Photo: Institute for Marine Geology, C.N.R. 1970.

Plate 5
Plane-parallel bedding: marker bed

Another panoramic view of a turbidite formation, comparable with that shown in plate 1, is proposed here. This is the largest and best-exposed "flysch" body of the Apennines, the Marnoso-arenacea Formation of Miocene age. Fluvial erosion and rejuvenation (repeated uplift) of the chain made these extensive outcrops, which are observed from some shorter distance with respect to plate 1. Bedding is marked by a vertical alternation of sandstone and mudstone beds. The more resistant sandstone is protruding from cliff faces, and some individual beds can be recognized. One, in particular (see *a*, and inset), 5 m thick and better cemented than others, stands out; it can be easily mapped as a *key bed*, for the whole formation (over a distance of more than 120 km). A key, or marker bed, is used as a stratigraphic tracer for correlating distant and separated parts of the same formation or succession: as turbidites accumulate very fast, each bed marks a geologic instant, a *time line*.

The sandstone of this particular bed was quarried to make building and paving stones; it is cemented by calcium carbonate, and locally called "*pietra serena*" or "*alberese*." The accompanying light band is a 7 m thick mudstone that completes the layer, deposited by a huge turbidite flow carrying more than 30 cubic kilometers of suspended particles. This was a highly catastrophic event, probably triggered by a submarine earthquake of large magnitude; a modern analog is represented by the 1929 Grand Banks turbidite in the Atlantic Ocean.

As a last observation, note that the strata are inclined; this is a secondary, or tectonic attitude because they were originally accumulated on the horizontal floor of a deep-sea plain. They were tilted around a horizontal axis that is called *strike*.

The two geologic formations illustrated here and in plate 1 have many features in common. First of all, they belong to the category of *flysch* units, bodies of detrital sediments deposited in strongly subsiding "mobile belts" of Earth crust. From these belts, narrow and elongate, and parallel to the margins of continents, mountain chains emerge on a time scale of several million or some tens of million years. Sedimentation is part of a complex sequence of geologic events, the so-called *orogenic cycle*, that culminates in the surrection and erosion of the mountains. During these vicissitudes, the sedimentary strata and the basins that accommodate them (they were once called *geosynclines*) are deformed by folding, shearing, and faulting. The deformation is related to subduction of lithosphere beneath the continental margin or to a collision between continental masses, and causes mainly compression and shortening of the crust. However, stratified slabs of various sizes, such as those pictured in these plates, can remain relatively undisturbed and display the original stratigraphy. The slabs form *tectonic units* (relatively rigid slices, sheets and shingles called *thrusts*) that are juxtaposed and superposed by the tectonic compression in mountain belts; surfaces of shear and friction separate them. They constitute tectonic (mechanical) boundaries, which must be distinguished from stratigraphic boundaries.

A second characteristic that is shared by the two formations is the basic sedimentary process, the turbidity current. This is not continuous, but punctuated in time as all catastrophic, or sporadic events. I introduced this mechanism to explain the lacustrine *varves* seen in plate 3, but the volumes of sediment involved in marine *flysch* basins are much larger. Thick and very thick beds, reaching volumes of 1 Km3 or more, are not uncommon; their estimated time of recurrence is in the order of thousands to tens of thousand years. Related beds and layers are often referred to as *meg-abeds* and *megalayers,* respectively. The one emphasized in plate 5 is named Contessa Bed. Single sedimentary events of such a huge size cannot be simply the result of fluvial floods or sea storms, though of large scale. A large reservoir of sediment, already accumulated somewhere, had to be available for instantaneous remobilization into deep water. The search for these "parking areas" or sediment repositories along basin margins is one of targets of sedimentological research, and a challenge for actualistic principles. Various areas have been recognized as "prone" in modern settings, but few have provided positive evidence for major events: the Mississippi delta is one of them. It has fed large turbiditic flows into the Gulf of Mexico, especially during glacial epochs, when sea level stood low and basin margins were more unstable. The pull of gravity could then induce great failures and sliding.

Photo: G. Piacentini 1970.

Plate 6
Parallel stratification: bedsets

We are getting closer to one of the outcrops of the previous plate. Individual turbidite layers (and particularly sandstone beds) are now well recognizable when their thickness is at least 50 cm. The presence of only two rock types and their repetitiveness have convinced some geologists to qualify successions of this type as *monotonous*. Sedimentologists are not so happy with this term, which suggests uniformity and boredom (rhythmical is more appropriate: see plate 3). Lithologic variety is not needed to provide an interesting stratigraphy; an educated eye can read various motives in so-called monotonous successions, too. Here, for example, sandstone beds are vertically separated by a variable thickness of mudstone; some of them are isolated within the mudstone, some are grouped in packets or *sets,* within which there is little mudstone. As we have seen in the Introduction (see figure 10), these bedsets can be called sequences if the thickest beds are concentrated near the base, the top or the middle. Within the two bedsets visible near the summit of the cliff, no apparent sequential organization is present, while one at the base displays a thinning up trend.

What is the meaning of these packets and sequences? Do they reflect the normal functioning of the basin or an alteration of it? What is seen is a change from conditions more favorable to sand accumulation than to conditions of prevailing mud deposition. This change did not occur on a layer by layer basis but from a multilayer interval to another multilayer interval. In other words, it had not the periodicity, or time scale, of individual sedimentation events but a wider one. Apart from this change of scale, two alternative hypotheses can be set up, as done in the case of plate 3:

1. The mud-rich parts of the section represent a "normal" or background sedimentation, in this case a "normal" frequency and size of catastrophic events producing turbidity currents. Occasionally, more voluminous currents, triggered by more violent events, brought large volumes of sand to the basin. When these paroxysmal bursts exhausted their sediment repository or their energy sources, calmer conditions came back and muddier turbidites of smaller size buried the sand-rich bodies;

2. The depositional conditions alternated between two states, one promoting and the other damping the sand input (a sort of chaotic behavior, in mathematical terms).

Sedimentologists do not know the answer yet: the two hypotheses have not been thoroughly tested, particularly by quantitative modeling (this method has been applied successfully in other branches of geology, not so in sedimentology and facies analysis). Geologists can expect, however, that the sand accumulated along basin margins in times of high sea level (when space was available in deltas, estuaries and shelves), to be remobilized and resedimented afterwards, when sea level dropped. Muddier intervals would thus reflect periods when sea level rose (*transgression*) or stabilized at high stands, whereas sand sheets would coincide with low stands, when the basin margins emerged (*regression*), and were subject to erosion and sliding (the two processes can be comprehensively called *denudation*). Cycles of marine transgression and regression are triggered by relative sea level ups and downs in the order of a few meters up to 100–200 meters, which have various causes and time scales. It was long thought that such cycles affected sedimentation in coastal and nearshore areas only; only recently it has been realized that they leave a mark in deep water, too.

Coming back to the descriptive aspects of the outcrop, two more points are stressed: 1) The vertical variations of "sandiness" can be expressed by a sand/mud (or sandstone/mudstone) ratio, i.e., the aggregate thickness of sand versus the aggregate thickness of mud for every interval of, say, 10 layers or 10 meters (or other predefined and fixed range). This index can be plotted in a diagram to emphasize peaks and fluctuations; 2) In sand-dominated bedsets, thinner mudstone beds are intercalated; thin sand beds are similarly interbedded within mudstone dominated intervals. In general, these relatively thin units of subordinate lithology within a dominant one are called *interbeds*, or *partings*.

Marnoso-arenacea Formation, northern Apennines, Santerno valley.

Plate 7
Meso-scale view of parallel bedding: turbidites and saw-tooth profile

There are many roads crossing the Apennines, and the turbidite layers of Marnoso-arenacea Formation can be approached along them for a closer examination. You see here an example, with the typical saw-tooth profile of sandstone beds, induced by fracturing and weathering of the rock. Two fracture systems, both perpendicular to bedding, isolate blocks of sandstone that periodically fall from cliffs and roadcuts. In this section, sandstone beds are relatively rare and separated by thick mudstone intervals. The mudstone, anyway, is not monogenic: three different types are present in this outcrop: a light, calcareous mudstone (a marl) forms a prominent bed in the central part; above and below it, alternating bands of different gray tone occur. They are a few centimeters thick and have a variable carbonate content (higher in lighter beds). A thin sole of cemented carbonate sand (calcarenite) supports the lighter marl bed.

On the whole, five lithotypes (sandstone, calcarenite, 3 types of mudstone) appear here. Let us see whether they form a random succession of beds or are associated into layers. Sandstones form couplets with darker mudstones (upper left- and lower right-hand sides), so defining turbidite layers. The calcarenite forms a similar couplet with the light marlstone: this is also a turbidite but of different composition (carbonate clastic) and provenance: the carbonate derives from the accumulation of small skeletons of organisms, whereas the predominant turbidite layers were fed from a siliciclastic, terrigenous source. Siliciclastic means a detritus composed of quartz and other minerals rich in silica, i.e., silicates; terrigenous means coming from land, or *terra firma*.

We thus have a calcareous turbidite interbedded with terrigenous turbidites. But the story does not end here. Dark mudstones, decoupled from sandstone beds, alternate with lighter ones; they represent the deposit of slow, dilute turbidity currents that carried no sand in suspension or had lost it in a previous tract of their path, closer to the source (see sketch attached to plate 69). The lighter mudstones are richer in carbonate remains of plankton falling from superficial waters; they mixed with terrigenous mud forming a *hemipelagic* deposit, i.e., the normal sedimentation of the periods when turbidity currents were inactive. These hemipelagic muds had a much slower sedimentation rate than turbiditic muds; for this reason the skeletal remains, representing the biological productivity of the ocean, reached a higher concentration there. As turbidites are practically instantaneous events, the actual geologic time is recorded by hemipelagites; the distinction between the two types of mudstone is thus relevant for dating the whole formation or parts of it: samples must not be taken from turbidite mudstones, also because they are almost barren, i.e., devoid of fossils.

In conclusion, two types (and ranks) of couplets can be defined here: one is formed by the two members of a turbidite event (sand + mud), the other by the alternation of a turbiditic layer and a hemipelagic bed. In this second case, the sand bed of the turbidite can be missing, leaving a muddy turbidite.

Marnoso-arenacea Formation, northern Apennines.

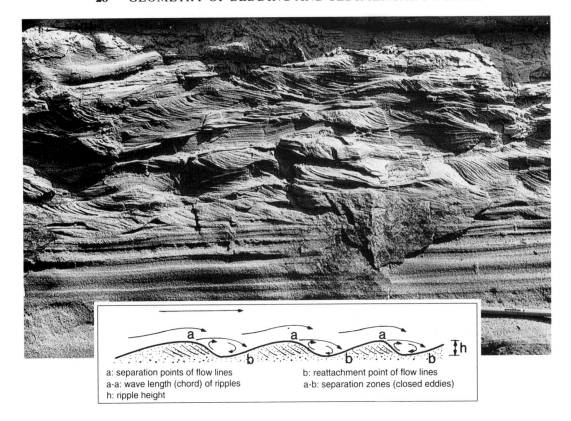

a: separation points of flow lines
a-a: wave length (chord) of ripples
h: ripple height

b: reattachment point of flow lines
a-b: separation zones (closed eddies)

Close-up of a bed: internal structures

Sedimentary structures can be easily discerned, up to the individual lamina, when looking at beds from a close distance. Laminae are depositional units and, at the same time, sedimentary structures. Their spatial aggregation and arrangement creates, in fact, peculiar geometries and morphologies within a bed. In this case, observe the passage from thick, plane-parallel lamination in the lower part of the bed, to asymmetrical cross-lamination in the upper part, where the laminae become also thinner. The parallel laminae indicate the vertical growth of a flat depositional surface where sand was supplied; the cross laminae do not interrupt this growth but mark a change from a flat to a rippled bottom. The ripples migrated down flow (to the right) and were continually buried under the accumulating sand and reformed on top of it. The sand was dragged along the bottom in both cases, before coming to rest; a *tractive current* was responsible for that. Tractive structures will be examined in a special section; what is remarked here, is the fact that the vertical succession of the two types of lamination is not haphazard. It indicates that the current, while forming these structures, was decelerating. How can one say that? If you look at the sediment grain size, you will note it is coarser near the bottom, then fines up (see table 2). The size of particles is function of the current velocity, so it is clear that the current slowed down during deposition. When the speed was still relatively high, ripples could not form; only a flat bottom was hydrodinamically stable. Hydraulic sorting of grains in the bed load grouped them in strips and "carpets" moving at slightly different veloc-

ity because of differential friction. Distinct laminae thus formed, each showing a particular textural signature.

When ripples start to grow, the bottom exerts a greater friction on the current and the mechanism of sand traction is modified: the grains climb the upcurrent (stoss) side of ripples, then slide down the slipface forming the lee side. Foreset laminae accrete down flow and cause the ripple to grow laterally. The sand grains are then buried by the newcomers and stop for a while, until the migration of the ripple form (with the stoss side bypassing the position previously occupied by the lee side) exhumes and entrain them again (see inset). This mechanism implies that the sand grains alternate motion with immobility, with the result that their average velocity is decreased in comparison with the flat bed condition.

The bed shown in the photo is again a turbidite (we linger in this world to stress both the interrelation of several aspects and the optimal display of various features of more general interest). The order of superposition of the two types of laminae is part of what is called a *Bouma sequence*, from the name of the sedimentologist who formalized it in 1962 (see more in plate 69). The same sequence, though with less detail, was recognized in the Thirties by geologists working in the Apennines, and used as a way-up criterion in tectonically complex settings, where the beds are frequently vertical (see plate 9) or overturned.

Marnoso-arenacea Formation, Sarsina, northern Apennines.

Plate 9
Verticalized parallel beds

When sedimentary beds are verticalized like this by tectonic stresses, good opportunities are created by differential erosion for observing structures on bed surfaces (see plates 82 to 103, in particular 88, 91, 98, 99, and 103), both in place and in detached slabs. The close alternation of softer and harder rock types is especially favorable in this respect.

This spectacular thin-bedded succession is made of: turbiditic sandstone beds (intermediate color and resistance to weathering), turbiditic mudstone beds (darker color, lesser resistance) and hemipelagic, marly limestone beds (whitish color, maximum resistance). Their mutual vertical relationships are those already described in plate 7, indicating the same depositional motives although in a different formation (Tertiary Flysch of Guipuzcoa, Spain). In spite of their modest thickness, the beds are remarkably continuous and sheetlike, as the pages of a book. They accumulated on a smooth, originally horizontal surface, similar to that of Modern deep sea plains;

turbidity currents exhaust their energy there, abandoning their residual load of fine particles.

How did these horizontal planes become tilted like this? In particular, did they rotate clockwise or anti clockwise around their strike? The presence of interfacial structures allow us to give the right answer to this question. Look at up facing surfaces of bedding: if you find ripple marks on them, for example, they represent bed tops (compare with plates 3 and 8). If, on the other hand, you see sole markings, you are looking at original bottoms. Sole marks, in fact, are preserved at the base of beds, and belong for this reason to *basal structures* (see figure 5 in the introduction). In conclusion, if the strata shown here are still upright in spite of their high inclination, they become younger to the left and have rotated anti clockwise. If they are upside down, the rotation occurred in the opposite sense.

Zumaya, near San Sebastian, NE Spain.

Plate 10
Lenticular sedimentary bodies

Vertical cliffs interrupt the milder slopes of this landscape in the Spanish Pyrenees. Outcrops of conglomerate (cemented gravel) make them. Conglomerate bodies are more resistant to erosion than the rocks they are interbedded with (clays and sandstones). Their lateral variation of thickness (wedging) can thus be appreciated. When wedging is observed on one side only, the body is called a *tongue* or a *wedge;* a *lens* tapers out on both sides (we are speaking of sections, i.e., bidimensional views). The lateral closure of a lenticular or wedge-shaped body, where base and top converge, is the *termination.* A depositional termination (there is no truncation above) is a

pinch-out. If the body splits laterally into several tongues, it *interfingers,* or *intertongues,* with an adjacent body.

Sedimentary bodies are made of several beds, which are not always apparent, especially in the case of sandstones and conglomerates, because of uniform cementation and vertical "amalgamation" (see Introduction, figure 8 **C**). The geometry of individual beds may conform or not to that of the enclosing body (examples will be shown in plates 11 and 12).

The lens-shaped conglomerates of this picture represent fossilized alluvial fans, fluvial sedimentary bodies accumulated at the termini of mountain valleys in Tertiary

(Oligocene) time. Torrential streams excavated these valleys and carried coarse sized detritus to a main valley or a plain located in a wide subsiding basin. Fans derive their name by the lateral expansion of stream currents flowing from mountain valleys and the radial shifting of channels.

Exposures like this, occurring near Pobla de Segur, Spain, are rather exceptional: the lack of tectonic deformation, the scarcity of vegetation, and the strong cementation of some rock types make them possible. The road at the bottom of the valley gives an approximate idea of the scale, which is comparable with that of a seismic section.

Plate 11
Sedimentary bodies: sandstone tongue and onlap

A set of sandstone beds pinches out in a mudstone sequence, barely visible on the right. As indicated by morphology, mudstones lie also on top of this tongue. When dealing with wedge- or lens-shaped bodies, a question to be resolved is the following: were they relieves that bulged up from an Ancient flat surface, or the filling of a depression such as a channel or a valley? It is implied, in the first case, that the base of the body was horizontal, and the top convex up; in the second, that the base was concave up, possibly owing to erosion, and the top horizontal.

The solution consists in finding the regional bedding attitude and suitable guidelines (*datum planes*) for paleohorizontality. Here, the sequence is slightly dipping to the left. The sandstone body is made of parallel beds, which are also parallel to the upper boundary of the set; instead, they make an angular contact with the basal surface of the tongue. The top is also parallel to the general attitude. Therefore, it can be assumed that it coincides with a paleohorizontal surface. If this is true, the beds abutting against the inclined base are said to lap on it, or to make an *onlap* contact (compare with downlap in plate 12). In other words, they are less inclined than the surface on which they rest. The conclusion is, that horizontal beds filled a trough by vertical accretion.

Marnoso-arenacea Formation, Santerno Valley, northern Apennines.

Plate 12
Sedimentary bodies: base-of-slope tongues and downlap

This natural section was cut by waves in a volcanic island of the Eolian Archipelago in the Tyrrhenian Sea. The bedded deposits were emplaced by sedimentary processes but are not sedimentary in origin; they are pyroclastic, and were ejected by explosive eruptions. Ashes made of volcanic glass, crystals and fragments of minerals solidified before the eruption, and blocks of indurated volcanic rock tore apart from the conduit walls, are all pyroclastic materials. In this case, they fill a crater, partly visible in the background. Half of this crater, later on, subsided below the sea along a fault, or was simply eaten up by marine erosion under the attack of waves. Thus, in the cross-section, the pyroclastic beds leaning against the vertical wall of the crater on the right. This contact is a high-angle onlap.

The deposition of pyroclastic materials may have occurred in two ways: direct or delayed. *Direct* deposition means that pyroclastic products are found where they touched the ground for the first time, after a ballistic trajectory and a free *fall* or the entrainment by hot, dense

suspensions flowing down the flanks of the crater (pyro-clastic flows). In *delayed* deposition, the particles have lost their heat: they fall after a long residence in the atmosphere or in seawater, being distributed over wide areas, or accumulate in the vicinity of the conduit but are remobilized afterwards. The final emplacement is thus delayed for two different reasons: 1) finer ash is pushed so high by rising, convection-driven gas columns as to reach the upper troposphere or the stratosphere and be involved in global circulation; it can stay there for years before returning to the ground with precipitation; 2) coarser and heavier particles, such as those forming the beds in the photo (see close-up in plate 69), are removed from the place of direct emplacement. The remobilizing agent can be rain, running water, wind, or simply gravity. Water-soaked ash is easily mobilized by gravity on steep volcano slopes and moves as a debris flow (which, in this specific case, is called *lahar*).

Sedimentological criteria are useful for recognizing di-rect versus delayed deposition. Among them, the geome-try of bedding is important. Fall deposits, for example, mantle a substratum, however irregular, with parallel *drapes*. Lahar deposits, on the other hand, are *lenticular* owing to their more localized character: the detrital mass

flows as a viscous, cohesive substance that follows the underlying topography and fills depressions with elon-gated, tongue or lobe-shaped deposits. In this picture, it can be seen that the beds are not perfectly parallel, which suggests mass flows and remobilization. Furthermore, they can be grouped into two sets: a lower set, subhori-zontal, and an upper one, slightly inclined. The lower set rests on the crater bottom (not visible because it is below sea level), while the upper set unconformably overlies the lower one with a *tangential* contact (inclined beds that become gradually horizontal) called *downlap*. Both sets lap on the steep crater flank.

The upper bedset is obviously nourished by the de-nuded slope to the right and on the background, and forms a *base-of-slope* body (or a *toeset*), which is tongue-shaped in cross-section. Base-of-slope bodies are point sourced and cone-shaped in 3-D, but frequently coalesce laterally and form *aprons*. The more abrupt onlap of the lower bedset implies that the pyroclastics came down other, more distant slopes, then flowed along the crater bottom for a while before coming to rest.

Pleistocene pyroclastics, Salina, Eolian Islands, Tyrrhenian Sea.

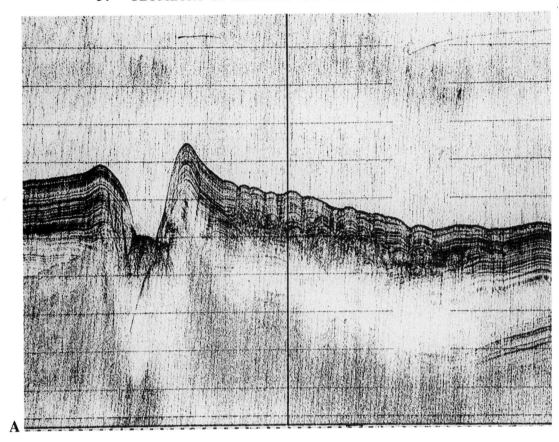

A

Plate 13
Erosional channel (A) and channel fill (B) in Recent sediments

These seismic sections show natural channels. One is still active (**A**), the other abandoned and filled by sediment (**B**).

The active channel is the submarine prolongation of the Crati River (Calabria, southern Italy) into the Gulf of Taranto, Ionian Sea. One may argue that this is the final reach of the river, submerged by the sea level rise that followed the melting of glaciers of the last glacial epoch. It occurs, however, at a depth greater than the rise in sea level (110–130 m); therefore, the channel was under water even during the glaciation and the low stand of sea level. Consequently, some submarine process is responsible for its excavation. The turbidity current has the proper requirements: when it gathers way and accelerates, it can acquire the power to erode the bottom. Like the discharges of river floods, the volume of turbidity currents may be quite variable. In more normal cases, the flow is confined within the channel but, sometimes, it is too voluminous and spills over. The part that flows out, with its sediment in suspension, decelerates because of expansion and lays down part of its load near the channel banks. Subaqueous levees are thus built up; they are wedge-shaped bodies that peter out away from the chan-

nel (see right-hand side of section). The growth of levees enlarges the channel section and increases its capacity; it can be asymmetrical, in which case the overspill preferentially occurs on the side where the levee is higher. This happens because of a deviation affecting all moving fluid bodies in a rotating planet (toward the right in the northern hemisphere). The deviation is known as the Coriolis effect.

The greater accumulation of sediment on one side of the channel can promote an instability of the levee slope, as illustrated here by deformed bedding; this is favored by the fine texture and high water content of levee deposits. The wrinkles you see in reflectors are produced by creeping or sliding of sediment; deeper reflectors are almost chaotic and then disappear (see white lens) indicating a strong absorption of acoustic energy sent in the subbottom (by a 3.5 kHz profiler).

In plate 13 **B,** acoustic waves show a deeper penetration (a different source is used, a 1 kJ Sparker): more reflectors appear, and well defined. Many are parallel and form the substratum of a distinct erosional channel filled by concave-up, onlapping strata (remember the vertical scale exaggeration causing distortion of shapes and

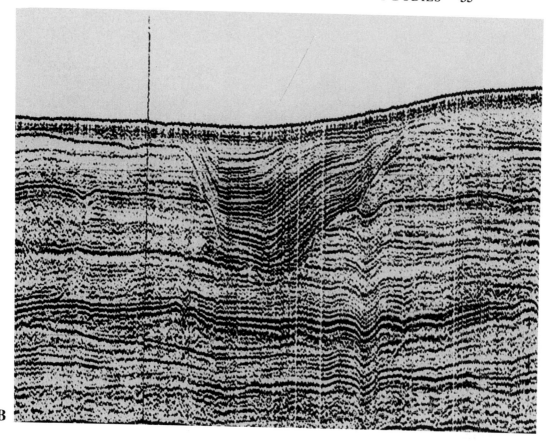

B

angles). A levee is partly recognizable on the right hand side by inclined, downlapping reflectors. The dark surficial drape is made of marine sediment, which transgressed an alluvial succession. This part of the Adriatic Sea was a prolongation of the Po Plain alluvial system when the sea level was lower, i.e., during the last Ice Age.

Other channel fill and levee deposits are buried in this late Pleistocene sequence: an example is the lenticular body in the lower left of the section.

Photos: Institute for Marine Geology, C.N.R., 1980, 1987.

The scale is lacking; as a reference, the channel in **A** is about 8 m deep, the channel fill in **B** is almost 15 m thick.

A

Plate 14
Clinostratification: prograding bodies

Clinostratification means primary inclination of bedding: the sediments were deposited on a sloping surface. This surface can dip in the direction of transport or across it; successive strata make a sedimentary body that grows frontally in the first case, laterally (transversally) in the second.

Frontal accretion is also called *progradation,* and creates dipping bedsets, made of shingling beds and called *foresets*. Foreset bedding can here be compared, at two different scales, in a wide outcrop of a gravel pit near Ancona (**A**) and a seismic section along the axis of the Adriatic Sea (**B**: south is to the left). For this growth model to occur, there must be an initial step, a difference in level between the point where the sediment is introduced and a deeper, horizontal bottom where gravity attracts it. This step migrates basinward by the more or less continuous addition of sediment. Deposition does not occur on its top, or is limited to thin sets of horizontal beds (*topset*), because room is not or scarcely available

there. A delta or a beach can advance into the sea in this way: sediment can be accommodated in front of them, below sea level, whereas vertical accretion is limited as it cannot occur above sea level. Aggradation is only possible, in coastal areas, when subsidence is active; room is then made by the sinking bottom, which is equivalent to a rising sea level, and sediment can compensate for it.

Foresets join tangentially the flat basinal deposits (see reflectors in **B**), also indicated as *bottomset*. In other cases, they make an angular contact (the already mentioned downlap). The landward connection with the topset can be either gradual (continuous deposition along the profile) or truncated by erosion; both cases are shown in the seismic profile, in the right and central parts, respectively. This upper contact is called *toplap*. Do not confuse topset and toplap: the former term indicates a mass, or volume of sediment, the latter one denotes a surface or, better, a contact between surfaces. All terms ending with "lap" are used for types of angular contacts, or uncon-

(M. Rainone, 1980.)

B

formities; to those already quoted, add here *offlap*, which is simply the combination of down- and top-lap.

A: The gravel shown in the picture (see also details in plates 62 and 64) was brought to the Adriatic Sea in the Pleistocene by a torrential stream similar to those draining the Apennines today. The pebbles, with some sand, were redistributed by waves and littoral currents along the shore, forming gravel beaches that dipped into the sea at relatively high angles. Similar beaches can be presently observed along the Ionian coasts of southern Italy. The pebbles rolled down the beach face under the effect of gravity; they thus escaped the mechanical action of smaller, fair weather waves, being removed only by strong storm waves. Normal waves were more effective on the beach top (swash zone and berm), where they had time to wash, abrade, and sort the pebbles. Horizontal topset deposits are, in fact, made of smaller, better sorted and more rounded pebbles than foreset beds; moreover, flat pebbles are here retained by solid friction as they have

a greater relative surface in comparison with spheroidal particles, more prone to rolling.

An even more impressive example of progradational body in outcrop is shown in color photo 2; it is a cross-section of a fan delta (a delta made of coarse clastics and leaning against a coastal relief), where foresets reach a thickness of several hundred meters.

B: During glacial epochs of the Quaternary, when land erosion was intensified and sea level lower, the Po delta advanced into the Adriatic Sea prolonging the alluvial plain into a mostly emerged shelf. The edge of this pro-grading shelf, now 140 m deep, is shown here, with the adjoining Mesoadriatic Depression, about 250 m deep. The youngest foreset deposits were truncated by wave and storm action when sea level rose and transgressed the shelf in Post-glacial times. Wave-reworked sediment formed thin topset deposits.

Photo: M. Rainone 1980; inset: Institute for Marine Geology 1982.

Plate 15
Clinostratification: laterally accreting bodies

A tabular body made of sandstone and conglomerate beds is here sandwiched between other similar bodies: its base runs close to the bottom of the photo, and the top is marked by an alignment of shrubs. The thickness is about 7 m. The section is of Pliocene age and crops out in a mountain cliff south of Bologna (the so-called Pliocene rampart).

At first sight, the inclined bedsets look like the foresets shown in plate 14. The main difference consists in the fact that they lie on an erosional surface, not on a depositional one. To explain how this erosion occurred, think of the lateral migration of a meandering fluvial channel (inset, right-hand side). In a meander loop, erosion takes place on the outer bank and on the channel bottom, deposition on the inner side. The sloping surface of the inner bank thus accretes toward the channel axis, forming a point bar. The bar consequently grows at an angle of 60°–90° with the main flow direction. The growth is not continuous but occurs mainly during floods, when the stream discharge is at a peak. The inclined bedding formed in this way is called *epsilon* bedding (it was labeled with this Greek letter in a classification of cross-bedding types of the 1960s, by J. R. L. Allen). The migrating point bar "runs after" the erosional bank in order to maintain a constant section for the flow in the channel; in doing so, it encroaches on the channel bed and buries it. The lateral migration of the channel bottom incrementally creates the planar erosion surface that will form the base of the point bar body. This surface is thus the cumulative result of many erosional events; in temporal terms, it is *diachronous,* or time-transgressive. In other cases, similar flat surfaces are, instead, the product of instantaneous, single-event erosion.

An important implication of this mechanism should be borne in mind: an erosional surface representing a fossil channel is not necessarily channel-shaped. If a channel changes its position, either by progressive migration or sudden shifts, the geometry that will be preserved in stratigraphic sections is not the instant channel form, but the record of shifting. This record consists in a flat erosional surface in the simplest case, as seen here, or in a more irregular and complicated one (with scours, steps, terraces, etc.).

Pliocene Intra-apenninic Basin, Zena Valley, northern Apennines.

Plate 16
Primary and secondary inclination of bedding compared

Plate 16 should clarify the distinction between clinostratification and tectonic dip of bedding, while the next one will insist on problems of interpretation of primary inclination. The outcrop shows fluvial and lacustrine deposits of late Miocene age in Tuscany.

All beds dip to the left but with different angles. Three bedsets are distinguishable with angular contacts between them. In the lower and upper set, sandstones and clays alternate in plane-parallel beds; the intermediate set is made of sandstones only. What happened? Were all sets deposited on horizontal planes and tilted afterwards at different times? Or were some inclined from the start, and, in this case, which ones?

Let us rotate the book clockwise until the lower set of the photo becomes horizontal. The situation then appears like that of the previous plate, with a set of inclined beds sandwiched between two horizontal sets. The depositional dip of the intermediate set is toward the right. Therefore, the paleo-dip is opposite to the present one, which has been determined by tectonic forces and is, therefore, sec-ondary. Now, the depositional setting can be interpreted with one of the models discussed before: frontal versus lateral accretion of a sedimentary body. If tectonic rotation had occurred in the opposite sense, i.e., clockwise, the primary dip of beds would have changed in steepness (increasing it) but not in direction.

When discussing vertical stratification (plate 9), I pointed out that tectonic tilting can occur clockwise or counterclockwise, and way-up criteria in beds can be applied to decide which the right way was. Here, the problem is compounded by that of recognizing a paleo-horizontal *datum* among various possible candidates. One can proceed by trial and error, manipulating a photograph until it assumes a convincing geometric setting. Convincing means *plausible* in terms of known sedimentological and stratigraphical models, such as those discussed so far. By the principle of parsimony, scientists give preference to the simpler explanation with respect to the more complicated one that comes to mind.

A

Plate 17
Inclined bedding: problems of interpretation

The tabular body standing out in the center of plate 17 **A** is about 20 m thick and composed of conglomerate (darker bands) and sandstone. It belongs to the same formation as the one seen in plate 15. The diagonal bedding indicates lateral accretion above an erosional surface cutting into marine mudstones. Medium-scale sedimentary structures (not visible) point to paleocurrent directions scattered within a wide angle, and are not conclusive for defining the main transport vector. The interpretation is thus open to both accretion models already discussed: a prograding coastal body (delta, beach) or a point bar in a shifting channel.

In actualistic terms, the basal erosion is best explained by a migrating channel. When looking at stratigraphy, however, one must take into account geologic time, particularly the time "hidden" by discontinuities such as an erosional surface. For example, deltas usually grow during high stands of sea level, with a submerged part made of marine sediments. If sea level rises rapidly, submerging a subaerial surface eroded during a previous regression, it is possible that the delta encroaches directly on this surface, because there is no time for deposition during the transgression (or the sediment supply is strongly reduced). The supposed deltaic body would have been of

the coarse type, called fan delta; in practice, an alluvial fan growing into a body of water. Such an apparatus has a small size and is composed of coarse materials because it is fed by a torrential stream cutting steep relieves. A fan delta displaying the "triad" topset-foreset-bottomset beds is also known as a *Gilbert type delta*. Its type model derives from Pleistocene deposits of Lake Bonneville in the United States.

Plate 17 **B** shows the uppermost part of a tabular body like that in **A.** The contact with the overlying sandstone body is erosional and underlined by a "pavement" of pebbles. The diagonal bedding is made evident by alternations of *interfingering* conglomerates and sandstones (conglomerate beds close to the left, sandstone beds to the right).

Pliocene Intra-apenninic Basin, Livergnano, northern Apennines.

This plate offers a chance for some methodological remarks. As said earlier, simpler and more "parsimonious" explanations are preferred in science. What is convenient for our brain, however, is not necessarily true; it should be considered as a first step in a series of successive approximations. In sedimentology, models of facies and sedimentary bodies

B

are built on some simplifying assumptions: that the supply of sediment is uniform and regular, that subsidence also occurs at a constant rate, and that sea level remains stationary. If, on this ground, we geologists obtain a reasonable and consistent picture, we are satisfied. If not, we try to transform constants into variables, one at a time, and see what happens to our model. During the whole history of our planet, the position of sea level changed many times, with variable frequency. As we lack absolute baselines, what we detect are *relative* changes of sea level, which leave their mark in sediments and in discontinuity surfaces. Some changes affect local areas or regions only, others are global: global changes are called *eustatic*, independently of their cause. Different factors, working at different time and space scales, can cause changes of sea level: for example, subsidence of a basin is a local factor; variation of ocean shape or volume is a global phenomenon induced by plate tectonics; the waxing and waning of glaciers reflects climatic forcing and is also global in scope.

The result of relative sea level changes are shifts of the shoreline across the continental margins. A *transgressive* pulsation means that the sea invades the land, a *regressive* one that the coastline advances seaward and enlarges the

land area. These opposite trends alternate in time and give rise to transgressive-regressive *cycles*. Both *during* transgressions and regressions, when sea level is unstable, erosion can prevail on deposition. In the first case, the erosional agent is the sea (through waves and storms); in the second case, subaerial agents like rivers, glaciers and wind are responsible. More or less extensive erosion surfaces can thus be formed.

After a rise or a fall, sea level stabilizes for a while; there are *high-stand* and *low-stand* conditions, respectively. During high stands, more space is available for accommodating sediment, especially on continental shelves and epicontinental seas (those invading the interior of continents). During low stands, this space is reduced and erosion favored. As a consequence, the progradation of a delta on a shelf is possible during a *relative* high stand.

The coarse-grained body of 17 A was deposited in a coastal environment, and could fit both the model of a prograding delta (in which case, the term foreset bedding is appropriate) and that of a laterally shifting fluvio-deltaic channel (epsilon bedding). If one is uncertain about the interpretation, the descriptive term *diagonal bedding* can be used.

Plate 18
Structureless beds (parallel)

Beds deposited by mass flow processes often lack internal structures and are welded on top of each other. As a result, bedding is barely detectable, and is described as *ill-defined;* or, you can say that the sediment looks poorly (faintly, crudely) bedded. The attribute "massive" is also used for this case, but should be avoided because it bears some ambiguity: not all "massive"-looking beds are produced by mass flows. Some are accumulated particle by particle (for instance, by fall of isolated stones from a wall).

Plate 18 shows pyroclastic beds made of coarse materials, i.e., cinder and lapilli. Cinder, or scoriae, are pieces of lava ejected from a volcano; they are hot, plastic and solidify soon after falling. At the same time, they weld together. Lapilli are pebble-sized, light and brittle fragments of lava that has already solidified but is rich of vugs (vesicles). Cinder and lapilli are emplaced near volcanic vents by fall or high-density mass flows driven by gravity.

The bedding is made visible here by changes in grain size and especially by the presence of thin drapes of fine ash (light lines).

The lack of structure within a bed can depend on two categories of causes: 1) the mechanism of deposition, and 2) a process occurring after deposition and canceling primary structures; such a process can be mechanical or biological (bioturbation).

In pyroclastic fall deposits, the coarsest particles are found at the base and closer to the vent, because they are heavier and land before the others (and travel shorter distances). In other terms, gravity operates a vertical sorting of clasts during fall, which results in an upward *grading*. The bed is not graded, it means that the fragments had small differences in size (weight) and fall velocity: one is probably very close to the vent. Most beds in the picture actually have an homogeneous texture.

Vertical grading is accompanied by lateral grading, with the average grain size decreasing away from the source; bed thickness also decreases.

Ejected particles that do not follow individual, ballistic trajectories may have another fate: they are so closely spaced that gravity entrains them in mass flows down the flanks of volcanoes. How can you detect this behavior? Look at the bed where the scale is placed. It displays an *inverted grading*, with coarser particles at or near the top. This cannot be due to gravity alone, i.e., to free fall. The particles must have moved in mass along a solid surface, bouncing and colliding owing to their high concentration. The effect of collisions within such a flowing layer is to expand the layer itself (you can try to reproduce this effect with Ping Pong balls): the largest particles are pushed farther from the bottom, while the finer ones sieve through them and approach the base of the flow. Mass flows of this type, sort of stony or *rock avalanches*, can move in a dry state if the slope is steep enough. The effect of water is that of lubricating the movement, whereby *wet* avalanches can occur on milder slope.

Plate 19
Structureless beds (lenticular)

In this plate, as in the previous one, coarse clastics prevail but consist of fluvial pebbles. Some lenses of sandstone, interbedded at various levels, emphasize the bedding, which is poorly defined where sandstone beds are absent (see center-left of picture). In a case like this, the vertical stratigraphy changes from place to place (imagine coring this section in different spots), and thickness measurements have a purely local value.

Even if you are not able to precisely define the geometry of every bed, the overall style can be described: the lack of parallelism implies a *lateral variability* of bed thickness, and the presence of closures, or terminations. These can be seen in the outcrop or reasonably extrapolated to its surroundings. In other terms, you can say that the bedding is *laterally discontinuous*. Remember that the concept of lateral discontinuity must not be confused with that of *vertical* discontinuity; the former is spatial, and expresses the localization or confinement of sedimentation; the latter is temporal, and indicates breaks in deposition.

The outcrop shows what could be an older equivalent of the deposits illustrated in figure 11 of the introduction, i.e., alluvial gravel with subordinate sand. The structureless beds were emplaced by mass flows during fluvial floods; flood events are fluid-driven, in contrast with the gravity-driven flows of the previous plate.

Pliocene Intra-apenninic Basin, Livergnano, northern Apennines.

Plate 20
Fine-grained structureless beds

Sediment of fine texture, whatever their composition, can form structureless beds, too. This example comes from the same Pliocene formation already shown in previous plates. It is exposed along a river bed, where the alluvial cover has been removed by recent erosion promoted by anthropic causes. The action of running water keep the rock surface "fresh and clean," whereas in normal outcrops it is usually covered by a weathered layer of variable thickness.

The thicker and softer beds are made of silt material (now compacted into siltstone). Various admixtures of sand and/or clay can be combined with silt. Siltstones and claystones are both comprised within the more general term mudstone. Thinner beds of sandstone are interspersed with the siltstone. Their top is wavy and the internal part laminated (back lighting masks the laminae). The hummocks were produced by waves (see more on this structure in plates 47–49).

The question, here, is to explain why unstructured and structured deposits alternate in time in the same place. Let us start with the sands, which do have structures; these structures indicate wave action, i.e., a relatively shallow water, where waves could exert traction and friction on the bottom. The supply of sand was modest and did not persist for a long time, thus giving thin beds. Assume now that the bottom deepens under the reach of waves, the so-called wave base. Waves are no more able to bring and mold sand. Finer material can settle on the quieter bottom. If it is supplied at a slow rate, it will take

a relatively long time to build a thick bed; if, on other hand, the sedimentation rate (quantity of sediment arriving on the bottom per unit time) is high, a thick bed can take a short time to accumulate. Whatever the rate, these beds could possibly reflect fluctuations of water depth around the wave base level. There is another possibility: that structures were present also in siltstone beds but were later effaced, for instance by organisms that thoroughly mixed and homogenized the sediment. From time to time, the sea bottom became inhospitable for these organisms, the waves sorted the sand grains out of the silty sediment, and the laminae could be preserved. Then, animal life repopulated the depositional interface and bioturbation started again. The two explanations are not necessarily incompatible, and might be combined.

In any case, what is remarkable in this reasoning is the application of the *principle of negative evidence,* which is of the utmost importance in geologic work. The absence of a character (structures in general, or specific types) does not mean *absence of information.* On the contrary, it conveys some message, first by excluding certain phenomena (those which produce the positive evidence) and, second, by obliging geologists to ask why a given character (which we would expect) is absent. Furthermore, there could be some hints at the cause of the absence; they could take the form of relics, or "ghosts" of the original structures or of overprinted features, such as traces of organic activity.   *Pliocene Intra-apenninic Basin, Reno Valley, northern Apennines.*

Plate 21
Stratification: hierarchy and cyclicity (1)

Facies similar to those shown in plate 20 alternate here again but on a different thickness scale. Sandstones are predominating over siltstones, and are organized in tabular bodies several meters thick. In the previous case, the alternating elements were individual beds; therefore, we have climbed a ladder in the hierarchical scale of rhythmic deposition.

Each sandstone body is composed of amalgamated beds. Wedging and undulations are visible in bed surfaces even at this distance. Bedding is not apparent, instead, in the lighter and more erodible siltstones.

Individual sandstone beds are vertically arranged in a roughly parallel way, but are lens and wedge-shaped. They have a limited lateral continuity, in contrast to the overall, tabular body. In other terms, the growth model is aggradational, as in the cases illustrated in plates 1, 2, 5, 6, but not of the layer cake type. This is because every single deposit affected a restricted area, and a lateral juxtaposition of beds was necessary to build up the whole body. The stratification can, in essence, be described in two ways: as a parallel, laterally discontinuous (wavy or wedge-shaped) bedding, or as low-angle cross-bedding. It depends on the definition of parallelism used (a broad or a restricted one). This geometry suggests deposition under the effect of strong storm waves, below sea level; the distance of observation prevents a closer examination of medium and small-scale structures, which would confirm

this hypothesis. The structureless siltstones mark phases of rising sea level and weakened water agitation.

Pliocene Intra-apenninic sandstones, Val Marecchia area, northern Apennines.

Where vertical changes of lithology and bedding style (or, more synthetically, of facies) occur, one should observe *how* they occur: abruptly or gradually, with or without discontinuity. A gradual transition from a fine to a coarse clastic facies commonly indicates a *shallowing* trend or the advancing of a body that has a depositional base; these processes characterize a *regressive* phase of sedimentation. An equally gradual passage from a coarse to a fine facies reflects a *deepening* or *transgressive* trend. In the example illustrated here, both the base and top of the sandstone bodies are sharply defined; therefore, neither of them matches the just mentioned cases.

What is certain, in any case, is that changes in depth and sediment input did occur. Even if changes follow a cyclical pattern as in the case of transgressive-regressive cycles, the fact is that these cycles are not always (or rarely) recorded by continuous sedimentation. Discontinuities occur rather commonly and are expressed by sharp contacts, like those seen in the picture. There is a hierarchy of stratigraphic discontinuities as there is a hierarchy of sedimentation units (laminae, beds, layers, bodies, etc.). The relative importance and the various ranks of surfaces of discontinuity and unconformities are not discussed here, anyway, because they

involve too many aspects of stratigraphy. Suffice it to point out the general meaning of discontinuities: when sedimentation stops and is replaced by erosion, a part of the stratigraphic record is canceled. If, for example, a shallowing trend culminates in the emergence of the sea bottom, the record of shallowing (a coarsening up sequence) can be entirely removed, and replaced by an erosional surface. Inasmuch as we sedimentologists are conscious that this can happen, we also know that the lacking sediments either were once present or never materialized. If we assume that they were deposited and then removed, we have a 50% probability of being right, and an equal probability of being wrong.

I add another remark: transgressive-regressive oscillations, strictly speaking, are horizontal shifts of the shoreline, not necessarily caused by vertical movements of sea level. *Local changes of depth* can merely represent the passive record of these back and forth movements of the coastline. In a particular point of observation, the water depth increases (or a transition between emerged and submerged conditions occurs) when the sea is transgressing. The opposite happens when the sea is retreating from land. A local phenomenon of subsidence, for example, favors a transgression by depressing the surface below sea level in a coastal area such as an alluvial plain or a delta. The compaction of buried sediments, or the extraction of fluids from them, can be sufficient to cause subsidence in a limited area. In the rest of the world, however, the level of the sea remains stable: there are no regional or global sea level changes.

Plate 22
Stratification: hierarchy and cyclicity (2)

We observe again, at a closer distance, a sandstone body comprised between poorly bedded siltstones: meso-scale structures are visible in this case. First of all, a vertical asymmetry can be noted: the base is sharpcut, the top transitional; component beds are thicker and coarser grained in the lower part, thinner and finer grained in the upper one. This package clearly shows a thinning and fining up trend. By the way, note here that vertical changes in grain size can occur at two distinct hierarchical levels: within individual beds or layers or within bedsets and multilayer bodies. To avoid confusion in descriptions, use different terms for the two cases: *grading* (or graded bedding) and *fining up*, respectively.

Further observations can be added to foster the interpretation of this sandstone body. The base seems to truncate underlying deposits and can be regarded as erosional. Above it, a structureless bed occurs, which could be interpreted either as a mass flow deposit or as a tractive deposit destructured by organisms or some other process. Then a set of roughly parallel beds follows, separated by wavy surfaces and internally subdivided into inclined (20°–25°) laminae dipping to the left. The lamination, of the foreset type (see plates 25–35), is the product of a tractive current that flowed from right to left and made subaqueous dunes on the bottom sediment. Finally, the topmost thin beds have the shape of flat wedges and contain laminae, which dip at much lower angles or are subhorizontal. As we shall see (plates 40, 41, 47, 48, 49), low-angle lamination is often produced by wave action.

Lower Miocene fossiliferous sandstones and siltstones of the Epiligurian Sequence, near San Leo, Marecchia Valley, northern Apennines.

This picture documents the temporal succession of different processes and events. The fossil remains (skeletal material within the sandstone) and the tractive structures testify to a marine, shallow-water setting. This is a broad interpretation, a first approximation. The sedimentary structures tell us more; why, for example, do we pass from current-dominated structures to wave-dominated structures? And what type of current made the cross-bedding? Moreover, we should try to understand the kind of basal erosion: was it produced by a single, instantaneous event, such as a big storm, or by many events distributed over a long time interval? In other words, was the erosion a catastrophic or a "normal," slow and gradual process?

We can imagine a shallow bay or a lagoon delimited seaward by barrier islands. The islands are made of sand and dissipate the incoming wave energy; they thus protect the embayment, where fine sediment can accumulate. Mechanical energy can, however, penetrate into the protected environment through channels between the islands (tidal inlets): these channels are cut and kept open by tidal currents to maintain water exchange with the open sea. The channels are not stationary if the waves impinge obliquely on the sand islands: some sand is deposited on one side, whereby the tidal current is obliged to erode the other side. In this way, the channel migrates as a fluvial meander does; its depth remaining constant, the results of this migration are the lateral prosecution of basal erosion and the progressive abandonment of the channel by lateral accretion.

With time, the channel bottom becomes a planar erosional surface; this surface is obviously diachronous. Sand is transported and locally accumulated in the lower section of the channel by tidal flows; dunes are formed in it and migrate in the current direction. The tidal motion is weaker in the upper channel section and outside the channel; there, waves leave their imprint on the sand. They form spits and

sand ridges, and sedimentary structures of various scales. Such features will be recorded in the topmost portion of the stratigraphic sequence.

In essence, this interpretation explains the sandstone body as the fill of a migrating channel by the combined action of tidal currents (main responsible for the channel existence) and waves in a littoral setting. The growth model would be a composition of vertical and lateral accretion. Physical evidence of lateral accretion (clinoforms) is, however, scarce, and other explanations are possible.

CHAPTER 2
Tractive Structures Produced by Currents and Waves

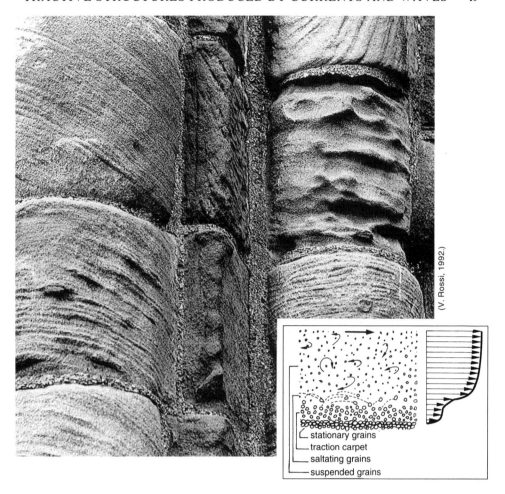

(V. Rossi, 1992.)

stationary grains
traction carpet
saltating grains
suspended grains

Plate 23
Tractive laminae

Sedimentary structures can be seen not only in the field but also in buildings and monuments, where natural stone blocks were employed. Various types of sandstone, for example, can be found in friezes, decorations, window-sills. Sometimes, sandstone is used for load-bearing elements, such as the columns in this picture, which are made of New Red Sandstone, a continental formation of Permo-Triassic age cropping out in the U.K. and Ireland.

In many cases the stone is chosen simply on the basis of local availability and economic convenience (cost of transport, labor, etc.). It is possible, however, to suppose that there was some aesthetic reason, too. One could be, for instance, the effects of light on the fine relief of the sandstone, as in the case illustrated here, where differential cementation and weathering emphasize lamination. Laminae in sand and sandstone are the most typical expression of tractive mechanisms, which are active in the bed load of a fluid-driven current. The bed load is formed by particles that move close to the bottom (see inset); due to their high concentration, the grains collide frequently.

The effects of the collisions, which increase with the momentum (current speed and particle mass), are twofold: sorting and abrasion. Like billiard balls, the grains move away after each impact and tend to join particles of similar characteristics (weight and size). The net result of

this "similar seeking similar" mechanism is to sort the grains and assemble them, during the movement, into relatively homogeneous arrays, which constitute the cores of laminae. At the beginning of traction, a few grains are mobilized but, with increasing current velocity, the whole bed becomes mobile and forms a *traction carpet*. The solid friction between moving grains and between them and the stationary bottom determines a resistance to transport; when this resistance exceeds the fluid stress, the grains stop, one string after the other, and a lamina is deposited.

A fresh supply of grains will accumulate new laminae on top of the previous ones. If the sand is deposited too rapidly, virtually in mass, the current has no time to sort the grains and form the laminae. Or, a faint lamination is only produced, and the water escaping upwards from the compacting sand deforms the laminae (as happened in the block on the upper right side of the picture).

Grain collisions also cause mechanical wear, manifested by the detachment of tiny chips that leads to rounding of corners and edges. Attrition is more intense in larger clasts, whereby pebbles are more rounded than accompanying sand for the same distance of travel.

Photo: V. Rossi 1992.

Plate 24
Bed forms: eolian dunes

Tractive structures are examined from two viewpoints: external shape and internal geometry. As it was said in the introductory section, the topography of both subaerial and subaqueous portions of the Earth's surface is related to sedimentary processes. In other terms, it coincides with the morphology of depositional surfaces; this can occur at various scales (see color photos 9–13), and is particularly obvious in the case of eolian dunes. Dunes and dunelike forms are the largest subaerial structures due to fluid traction.

In planform view, dune crests are more or less sinuous. Transversal dunes (color photo 9) have crests at about 90° to the main wind direction, and derive from the coalescence and rectification of barchan dunes (color photo 11).

Strongly sinuous, attached dunes seen from the ground look like waves in a stormy sea. The rear, upwind side is less inclined than the steep frontal (lee) side. The slipface of the lee side is rather smooth but a shallow relief is created by scars and grooves due to avalanching of sand grains. The sliding sand forms elongate lobes in the lower slipface. The stoss side of dunes is rippled. This difference is related to different mechanical conditions of the two surfaces (inset): the stoss side is subject to traction and friction by the wind. The sand grains are entrained until they reach the crest, then their behavior is controlled by gravity alone, because the wind loses contact with the ground beyond the crest edge, and a shadow zone (separation bubble) forms there. Shear is no longer felt on

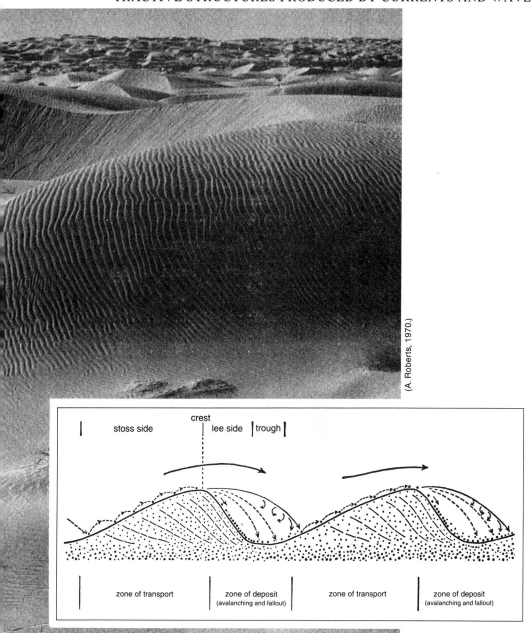

(A. Roberts, 1970.)

the lee side, and sand grains slide down the slipface when a critical angle is exceeded; this angle is called *angle of repose,* and represents an equilibrium between the pull of gravity on the slope and the internal friction of the sand which resists the movement. The internal friction, also expressed by an angle, is connected with grain shape and spatial arrangement (packing, fabric). In dry sand, this angle varies over a wide range (18°–35°); values of 40° can be exceeded when the sand is wetted by rain or dew.

The transport of sand on the stoss side of dunes, and its deposition on the lee side make the dune form migrate in the wind direction. Strictly speaking, only the back portions of dunes (and the ripples superposed on them) are tractive forms; actually, the fluid force "cooperates" with gravity in producing the whole bed form. Dunes grow from smaller relieves (sand mounds behind obstacles, foredunes, see color photo 12); the roughness they create on the ground surface and the transport of sand absorbs a lot of wind energy. The growth of vegetation

tends to stop the migration of dunes and to stabilize them.

Ripples represent the same process (friction at the air-sand interface) on a smaller scale. Some differences, however, exist. For example, the grains can jump from crest to crest when *saltation* is the dominant mode of movement. The impact of landing grains dislodges other grains, which are then mobilized in a sort of chain reaction. Ripples created by saltation are called *impact ripples,* in contrast to normal ripples where the grains climb the stoss side and slide down the lee side.

Dunes and ripple represent forms of *dynamic equilibrium* between flow and bottom sediment; this means that they are neutral in terms of both deposition and erosion. In other words, no net deposition or erosion occur during their formation: the sediment is in transit and the interface in a steady state. Some bed forms, however, are stable even in conditions of net sedimentation (see plates 54–56). *Photo: A. Roberts 1970.*

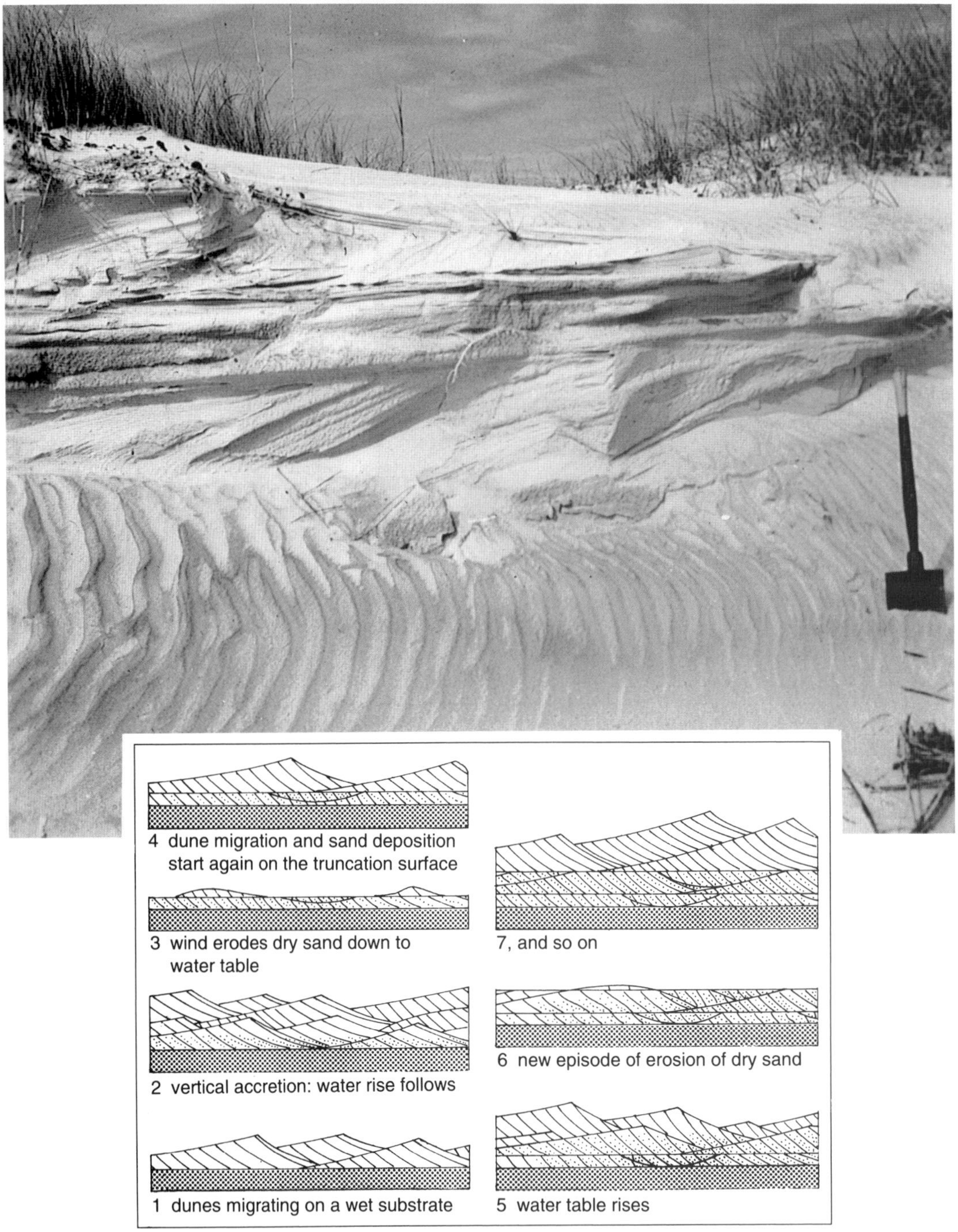

(A. Bosellini, 1992.)

4 dune migration and sand deposition
  start again on the truncation surface

3 wind erodes dry sand down to
  water table

2 vertical accretion: water rise follows

1 dunes migrating on a wet substrate

7, and so on

6 new episode of erosion of dry sand

5 water table rises

Plate 25
Eolian dunes: internal bedding

A trench cut in a dune along the wind direction shows the internal bedding (we see the upper portion only). In the lower part of the section, a set of dipping laminae (foreset laminae) reflects the leftward migration of the dune. Unconformably lying over it, are sub horizontal laminae and laminae dipping in opposite directions with low angles. The truncation of the foreset also represents the dune migration, in particular the passage of the stoss side, dominated by erosion or non deposition: the foreset laminae were removed, except for the lowermost part that remained buried in the ground. The low-angle, roughly parallel lamination could indicate *interdune* deposits. Interdune areas are relatively depressed zones intervening between dunes; they are occupied by a thin sand pavement and are sometimes invaded by water, in which case some mud can be interbedded with the sand.

Near the top of the section, a new foreset appears, again truncated by low-angle laminae dipping upwind. Low-angle laminae could also represent a limited episode of deposition on the stoss side of a dune.

Inset: modified from Collinson and Thompson 1982.

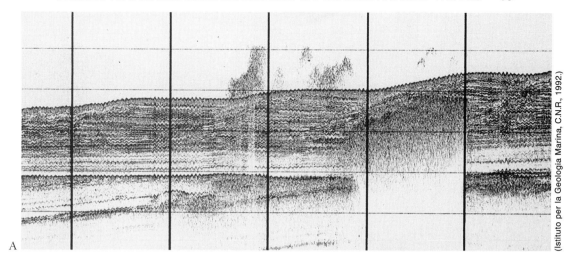

A

(Istituto per la Geologia Marina, C.N.R., 1992.)

B

(G. G. Ori, 1992.)

Another possible explanation of the abrupt truncation of foreset laminae lies in the presence of a water table (inset); in that case, water-saturated sand offers more resistance to wind erosion than dry sand, and is preferen-tially preserved. Exceptionally, giant foreset bedding is formed and preserved (see color photo 14).

Photo: A. Bosellini 1992.

Plate 26
Large-scale subaqueous bed forms (sand waves, sediment waves)

Isolated or repetitive, large-scale bed forms, similar to dunes in many respects, occur under water. They were first discovered by means of echo sounding devices, then "X-radiographed" by acoustic waves. The external profile of these structures was thus known before the internal geometry. The term *sand wave* is applied when their core is made of sand (**B** and **C**); if finer sediment prevails, as is the case in many seismic sections (**A**), the form is called a *sediment wave*.

A fundamental difference between sediment waves and classical dunes is the steepness of the slipface. In sedi-ment waves, the foreset slope is gentler and joins tangen-tially (asymptotically) the bottom. Laminae dipping as much as 20° can be found in sand waves (see **C**), but on the average their inclination is lower than the angle of repose. This means that the slipface is not protected from current shear or, in other terms, is not subjected to gravity only. Flow conditions and distribution of turbulence near the bottom must consequently be somewhat different in comparison with the eolian setting.

Moreover, rarely do laminae in foresets form a single, conformable set. Unconformities and changes of inclina-tion are visible in long exposures like those of pictures **B** and **C**. These features indicate that the migration of the form was not a continuous process: it continued for a while, then stopped and was reactivated when sediment was again available. For this reason, the truncations are called *reactivation surfaces*. During pauses in migration, erosion possibly occurred, especially in the upper parts of foresets.

The tangential character of the lower foreset (toeset) in sand waves is visible in the two uppermost beds of picture **C**.

A: *3.5 kHz subbottom profile offshore the Tiber delta, Tyrrhen-ian Sea. Spacing between horizontal lines is about 9 m; between*

C

(G. G. Ori, 1992.)

vertical lines, about 200 m.; **B, C:** *Greensand Formation, Cretaceous, U.K. Person for scale.*

Sediment and sand waves have a marked longitudinal asymmetry in both profile and internal bedding. This is the effect of their being produced, like eolian dunes, by a current flowing in a constant direction. But what kind of current produces large bed forms under water? In water masses, enormous amounts of energy are stored. The question is, how concentrated this energy is to do mechanical work. To build up sediment waves, large volumes of materials must be mobilized; furthermore, the sediment moves slowly, particle by particle as in all tractive structures. Consequently, the currents must persist for relatively long times to do their work. Episodic sedimentation and catastrophic events are not adequate: they are too short and cause mass transport.

Oceanographic surveys have found that currents with the necessary requisites can exist both in shallow and in deepwater: tides produce the most persistent and powerful currents in continental shelves (with velocities up to 1,5 m/s near the water surface), geostrophic currents do the same in cold, deeper water. Occasionally, strong winds blowing over long fetches can also produce bottom grazing currents in shallow water.

Most tides have a diurnal or semidiurnal period, and are accompanied by currents that reverse their flow for each semi-period (flood and ebb flows), and so last for some hours. Peaks of velocity (several decimeters per second) and carrying capacity occur for even shorter intervals. Peculiar morphological and oceanographic settings can, however, enhance the power of tidal flows.

Bottom currents related to the global, thermohaline circulation of the oceans, flow from the poles to the Equator. Especially where they find constrictions, their velocity is sufficient to remove sediment and build up sediment waves and drifts. As the currents follow the contour lines of continental margins, the sediment they accumulate is known as *contourite.* It can be pelagic, terrigenous or hemipelagic.

Photos: **A** Institute for Marine Geology 1992; **B** and **C** G. G. Ori 1992.

(G. Piacentini, 1970.)

Plate 27
Large-scale cross-bedding

This section, cropping out in the mini state of San Marino, included in northern Italy, is comparable with plate 26 **C,** and shows cemented calcareous sands made of skeletal debris (biocalcarenites) of Miocene age. The outcrop size is insufficient to show the whole geometry of the bed form that originated the cross-bedding. However, judging from the set thickness, in excess of 1 meter, we are dealing here with a large-scale form. Three unconformable bedsets can be discerned with variable dip direction; this should indicate that the section is not cut parallel to the paleocurrent (longitudinally) but at some angle (compare with plate 26C, where all foresets dip the same way).

Miocene "foramol" platforms of western Mediterranean, where many benthic organisms (Bryozoa, Echinids, Algae, *Fora*minifera, *Mol*luscs, etc.) lived in shoals and banks, were locally swept by strong tidal currents. These carbonate environments, characterizing temperate climatic zones, were later split by tectonic movements of the Apenninic orogen, and their pieces were displaced toward the foreland as *allochtonous slabs* floating on top of a peculiar thrust sheet (Ligurian).

Monte Fumaiolo Formation, Epiligurian Sequence, San Marino, northern Apennines.

In terms of scale, *subaqueous dunes* **must be distinguished from both eolian dunes and sediment waves. They are smaller, with a relief in the order of a few decimeters (below 5 cm, similar forms are regarded as ripples), and length in the order of decimeters or meters at most. In stratigraphic sections, it is convenient to use only one parameter,** *thickness,* **to define the scale. When you find a set of inclined laminae thinner than 5 cm, you are dealing with** *small-scale* **structures (ripples, ripple cross-lamination). When the set thickness (***normal to its bounding surfaces, not to foreset planes!***) is in the range 5–30 cm, you get** *medium scale,* **i.e., normal, dune-size cross-lamination (or cross-bedding). The upper limit is quite arbitrary, and others might be equally proposed (for example, 50 cm or 1 meter). Anyway, set thicknesses exceeding this limit define** *large-scale* **structures.** *Photo: G. Piacentini 1970.*

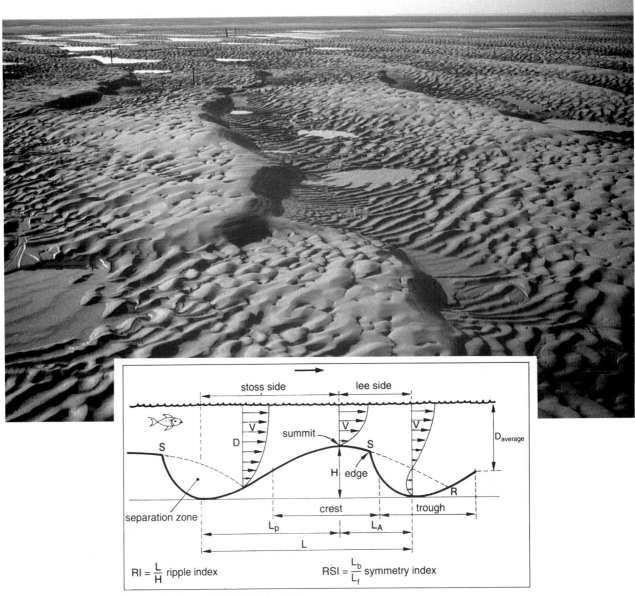

V = velocity; L = length; H = height; D = depth

Plate 28
Bed forms: subaqueous dunes and ripples

The expanse of sand reaching uninterruptedly to the sky-line is a view at low tide of a tidal flat, located along the SE coast of the North Sea (Oster Schelde, The Netherlands). At high tide, the same area is completely covered by water. The tidal flat is a depositional surface, where currents produced by tides dissipate their energy and transport sediment. These currents attain maximum velocity during rising or ebbing of the tidal wave, and zero velocity during high and low stands (slack water phase). In some parts of the flat, sand is the dominant sediment; in others, it is mud or an alternation of sand and mud in various proportions.

Bed forms of two scales (dunes and ripples, basically, plus some minor structures) characterize the morphology of this interface. They are produced when a water cover is present (and the current speed is at least 20 cm/s), and are preserved, with some modifications, when it retreats seaward. The implication is that the bedforms still exhibit their *equilibrium profiles*. Ripples stay on the back of dunes.

Both ripples and dunes are markedly asymmetrical, with a narrow and steep slipface; grain avalanching is the dominant process in subaqueous slipfaces as well as in subaerial ones of eolian forms. The asymmetry of the structures indicates that flood and ebb currents have not the same power, but one is clearly dominant over the other, simulating the effects of a purely unidirectional flow such as a river current. Only some details point to a tidal environment; among them, smoothing of crests by the opposite flow and reactivation surfaces (not visible

(G. G. Ori, 1992.)

here but only in section). The partial effacing and flattening of ripples along the dune crests, and some terracing of slipfaces, are effects of water shallowing and emergence, but can also occur in a fluvial bed.

In plan view, the crests of bed forms are highly sinuous with prevalence of concavities facing down current. This is equivalent to attaching, side by side, *lunate dunes,* the smaller subaqueous correspondent of desert barchans (see plate 30, inset).

The inset summarizes the geometrical parameters in the basic profile of subaqueous bed forms. Velocity vectors (arrows) are enveloped by vertical *velocity profiles*, whose gradient is related to friction with the bottom. The flow region influenced by bottom or side friction (i.e., close to a solid boundary, generally speaking) is called *boundary layer*. The flow lines in proximity of the boundary are skin friction lines.   *Photo: G. G. Ori 1992.*

Plate 29
Bed forms: current ripples

In this close-up, we can see the edge of a subaqueous dune and the adjacent trough where still water is trapped. The ripple morphology can also be appreciated. Ripples at the bottom of the pool are smaller and barely outlined; this is, in fact, an area protected from stronger eddies of the current (a shadow zone). The ripples are also arranged in a fanlike pattern (ripplefan).

Besides the flattening of ripples on the dune crest, we observe here two other types of *modification of the subaqueous forms due to shallowing* and emergence at low tide: 1) "microdeltas," i.e., small lobes of sand in-

cised by tiny channels (rills) which indicate draining of water and sand toward the residual pool (upper left); 2) small ripples superposed to normal ones, with crests at 90° (see upper right). They are named interference ripples (see also plates 50–53), and should not be confused with other minor structures parallel to the flow, such as the grooves and spurs, which are visible in the foreground and in plate 30. Grooves and spurs are part of the equilibrium morphology of subaqueous ripples.

Photo: G. G. Ori 1992.

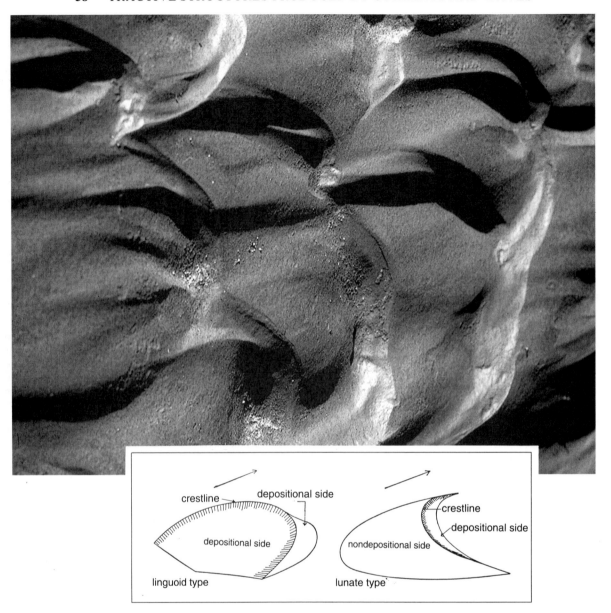

Plate 30
Bed forms: lunate ripples

Ripples whose profile changes little along crests are called linear (plate 31); they approach the hydrodynamic equilibrium profile for a bi-dimensional flow, a flow whose configuration does not change across the direction of movement. When, on the other hand, transversal perturbations, or "side effects," exist in the flow, the bed forms acquire a three-dimensional morphology: their profile changes laterally, as in this picture. Two basic "3D" shapes exist: *lunate* (or *barchanoid*), and *linguoid*, de-

pending on how the crest curvature is oriented with respect to flow direction (see inset). Actually, one can be misled by "3D" morphologies when trying to identifying and measuring this direction. Remember, then, that the safer *indicator* of the flow vector is the steeper side, which faces down current in any case.

The ripples are often ornamented by minor structures parallel to the current or fanning down current (spurs).

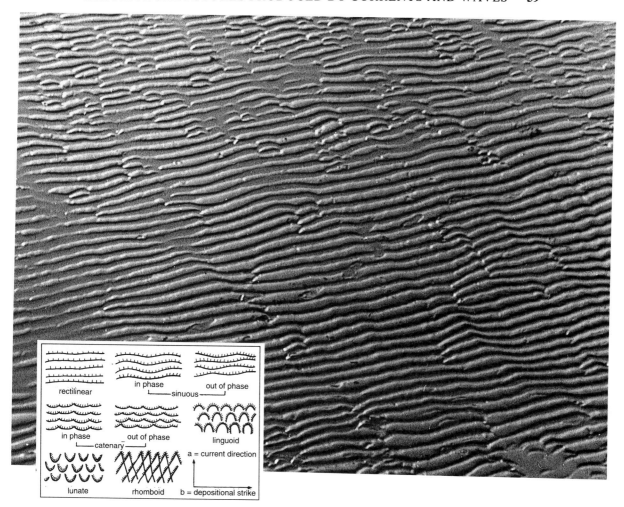

Plate 31
Bed forms: linear ripples

Rectilinear ripples have a good lateral continuity and parallelism, and are not usually associated with larger forms. They can be produced by both waves (as in the case illustrated here) and currents that are almost bidimensional, in the hydraulic sense. The flow lines develop secondary eddies near the bottom that are markedly cylindrical (rolls), and dovetail at their ends with similar neighbors creating the junctions you see in the picture.

The flat sandy bottom is that of a subaqueous beach exposed at low tide along the Adriatic shoreline. The ripples have flattened crests; they were formed and modified by waves in shallow water. A fossil analog (apart from the flattening) can be found in plate 45 for comparison.

Subaqueous ripples and dunes are equilibrium forms, as said before. When the physical conditions change, these forms become unstable and change into others or are canceled to fit the new conditions. However, they can have some inertia, or resistance to change, which depends on time or other factors. If the current velocity and the water depth on a dune-covered bed are slowly reduced, for example, dunes are actually smoothed out and replaced by ripples. Ripples, however, do not disappear even when the current has completely waned. Dunes can survive, as shown by plates 28 and 29, if the current decelerates and/or the water drains away in a relatively short time. As for the ripples, there is no problem of time: they persist in any case. One may thus conclude that ripples have the maximum preservation potential among bed forms; consequently, it is no surprise that they constitute one of the most common structure found in sedimentary rocks, considering also that weak currents are sufficient to form them, and that some sort of current can exist at any depth in any aqueous environment.

Bed forms not wholly effaced by changing conditions are altered or modified. *Modified structures* can be defined as the remnants of original structures, *modification structures* as the overprinted ones.   *Photo: G. Piacentini 1970.*

(Istituto per la Geologia Marina, C.N.R., 1992.)

Plate 32
Current ripples in deep water

This photo was shot by a subaqueous camera towed by a ship, at a depth of about 300 m in the Sicily Channel, Mediterranean Sea. There is no scale, but the coarse texture of the material (sand and small pebbles, probably) is evident. Asymmetrical ripples (or megaripples, which are roughly equivalent to dunes in size) were made by a current whose direction can be detected by the position of the slipface. Waves, on the other hand, are not capable of moving sand and pebbles at this depth.

In an inland sea like the Mediterranean, the bottom topography and water circulation are quite complex. There are different basins of diverse size and depth, which communicate by means of submarine sills and straits. Lateral and vertical confinement of water masses can occur, which induce local acceleration of currents; these currents are created by surface winds, pressure gradients, and water exchanges between basins with different hydrological balances (the eastern Mediterranean, for ex-

ample, is subject to strong evaporation and draws water from the western part). Tides and global circulation, on the other hand, are rather ineffective.

Whatever its origin, a current exerting a mechanical action on sediment on the deep-sea bed (by *deep* I mean "beyond the shelf edge") is called a *bottom current*, which is a convenient descriptive term. A bottom current can be sufficiently strong as to drag coarse particles; if it is weaker, only mud can be carried in suspension. A tractive bottom current is a typical selective agent; it separates sand from mud leaving beds of "clean," sorted to well-sorted sand. If terrigenous material is not available, the sand can be made of organic remains (calcareous or siliceous). Bottom currents generally flow along the contour lines of the sea bottom, and are also called *contour currents*. Their deposits, regardless of grain size, are consequently named *contourites* (see also plate 26).

Photo: Institute for Marine Geology, C.N.R. 1992.

Plate 33
Foreset laminae: the "basic" set

In vertical cuts, bed forms appear as cross-bedding or cross-lamination, as we have seen in some large-scale examples (sand waves, dunes). There is no unequivocal rule for using the term lamination instead of bedding or

stratification. The important thing is to specify the *scale*, in terms of length (in outcrops of sufficient extent) and thickness of the structure. Bed or laminaset thickness indicates the form height, coinciding with it if the whole

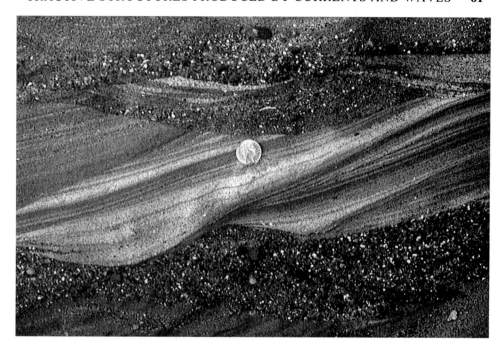

form has been preserved; otherwise, thickness gives only a minimum estimate of the original height. Local erosion of the bottom is an inherent aspect of the traction mechanism, and preferentially affects bed form crests.

Tractive features are formed under two conditions:

1. **Pure traction, with no net sedimentation.** The bottom does not aggrade, and bed forms simply migrate, canceling and replacing one another. As no new sediment is added, the current *reworks* a previous deposit and *molds* it.

2. **Traction accompanied by deposition.** The current slows down and loses carrying capacity but the hydraulic and morphological conditions remain in the stability field of specific bed forms; the bottom accretes, and the forms are progressively buried and remolded.

In the former case, illustrated by the picture, only one set of cross laminae is formed; in the latter, several *laminasets* are superposed with intervening discontinuities (local erosion surfaces: see plates 8, 54, and color photo 16). The delicate point is the identification of individual sets, representing single, continuous events of bed form migration. Every set is comprised between unconformable, or discordant surfaces: the coin staying for scale, for instance, is leaning against the middle of a laminaset made of sand, sandwiched between lenses of fine gravel. All the sand laminae dip to the left, but not at a uniform angle: from right to left, 4 packets can be distinguished. They are separated by angular contacts due to erosional truncations and representing stop-and-go of the bed form migration (if you remember, such surfaces are called *reactivation surfaces*). When measuring the set thickness, attention must be paid not to cross one or more such unconformities: for example, if one measures the thickness of the whole sand bed near the left side of the

picture, one will obtain a value that is almost double that of the individual set.

If the coset, or multiple set of laminae, was produced by the same type of bed form, it practically consists of a repetition of the same *basic unit,* or basic laminaset (see, for example, plates 6, 36, 54).

Recognizing the vertical expression of sedimentary structures is important, especially when dealing with samples of limited lateral extent, such as hand specimens and cores. In the case of cross-bedding, care must be taken in estimating not only height of the structure (through identification of the basic set, see above), but also the inclination of foreset laminae.

This angle is referred to the base of the bed and reaches its maximum value in sections cut parallel to the paleocurrent. If it roughly corresponds to the angle of repose (which is known for various types of natural grains), the structure is a dune or a ripple. If, on the other hand, it is smaller, other bed forms are implied (see plates 37–41). The two cases are termed, respectively, *high-angle* and *low-angle cross-lamination*, and indicate different *hydraulic regimes* (combinations of current or wave velocity and water depth).

In doing this important test, one must realize that a random section of the sediment can show either the real or *apparent* inclination of the dipping laminae. A section making an angle of 45–90° with the paleocurrent direction will be strongly oblique or normal to it, and will show foreset laminae inclined at angles smaller than 15° even if the actual inclination is 30°. It would be convenient to get several sections, with different orientations, of the same bed, and compare them. The highest recorded angle can be taken as an approximation of the true dip angle, and that section will be assumed as about parallel to the flow.

To recognize the proper orientation of sections with respect to paleocurrents, see plate 34.

(G. G. Ori, 1992.)

(G. G. Ori, 1992.)

Plate 34
Trough cross-bedding: transversal (A) and longitudinal (B) sections

A *longitudinal* section is normal to master bedding and parallel to the current direction. Laminae are seen dipping down current at the angle of repose of sand. Three laminasets, corresponding to as many beds, are visible plate 34 **A;** the handle of a shovel stands for scale at the bottom. The thickness of the central bed, which is truncated atop, is more than one meter; this indicates that the original bed form was larger than a dune, and should be called a sand wave.

What kind of current made it? Here we are observing Recent sands along the Dutch coast of the North Sea, where strong tidal currents are active. Therefore, it seems logical to assume that similar currents were responsible for originating these large tractive structures. A peculiar *indicator* of tidal influence, moreover, is present and consists of dark bands recurring rather regularly within the foreset. These bands are mud drapes rich in organic matter, marking pauses of slipface advance; similar but thinner mud laminae are present (but not easily visible) within each laminaset. In high-energy tidal environments, mud settles during slack water time, at high and low tide stands, when the current inverts its direction, and its

strength becomes insufficient to transport sand. The tide range and the ensuing power of tidal currents vary during a monthly tidal cycle (spring-neap cycle), minimum values being recorded at neap tides, every 14.5 days (half lunar month). These minimums are matched by enhanced slack water effects, resulting in thicker mud drapes.

Two downlap and as many toplap contacts can also be observed in this picture. The toe of the intermediate bed shows an alternation of sand and mud laminae separated by wavy surfaces. These materials accumulated in a trough between sand waves (compare with plate 29, allowing for the different scale). The sand was rippled by waves and draped by mud in slack water phases.

A *transversal* section contains the vertical direction and the orthogonal to the current. In 34 **B**, the paleocurrent was at right angles with the page. The shape shown by cross-bedding is that of *troughs* or *festoons,* concave-up laminasets separated by scour surfaces (see drawing). This profile can vary, however, with the three-dimensional geometry of the generating bed forms. Forms with linear crests, for example, give transversal sections made of parallel laminae, plane or slightly wavy. Trough cross-

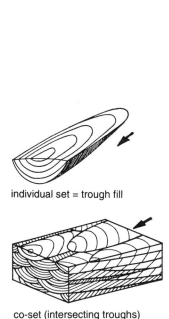

individual set = trough fill

co-set (intersecting troughs)

bedding reflects markedly three-dimensional, elongated ripples or dunes, whose tops are eroded, while the fill of troughs is preferentially preserved.

Every set in picture **B** is some meter thick, and represents a phase of migration of Ancient eolian dunes belonging to the Navajo Sandstone, a Jurassic formation of the western U.S. A longitudinal section of the same unit can be seen for comparison in color plate 18. Note that laminae dip in one direction only in sections parallel to flow, in opposite directions in transversal sections. The same criteria for discriminating between longitudinal and transversal sections apply regardless of scale, from sand waves to ripples.   *Photos: G. G. Ori 1992.*

Plate 35
Trough cross-bedding: "horizontal" section

Among the three characteristic sections of cross-bedding, which are mutually perpendicular (see inset), one is parallel to the topographic surface (i.e., horizontal in Modern settings: see commentary below). It contains the current direction (usually indicated by x or a in a spatial frame of three cartesian axes) and the normal on the horizontal plane (z or b). The current vector coincides (see arrow in sketch, hammer in picture) with the bisectrix of the arc outlined by the intersection of foreset laminae with the horizontal plane (this pattern is also called a rib-and-furrow structure, see plate 94). Compare the sketch with the picture and find out the paleocurrent direction in the outcrop.

The width of the trough (and of the corresponding laminaset) is measured along z.

A horizontal or bed-parallel section is not common; when available, it provides the best opportunities for paleocurrent measurements. This for two reasons: 1) the single reading is more reliable because the geometry of the structure is more evident in this section; 2) multiple measurements can be taken in troughs juxtaposed at the same stratigraphic level.

Pliocene sandstone, Intra-apenninic Basin, Brento, northern Apennines.

To describe the geometry of cross-stratification, reference has been made to the horizontal plane (an approximation of the present topography, at least in depositional environments) and to the vertical or zenith. In tectonically tilted successions, the terms horizontal and vertical must be replaced by *parallel* and *normal* to bedding planes, respectively. Reference bedding planes are, in certain cases, ideal surfaces that must be inferred by "averaging" inclined surfaces (for example, plate 34 A) or looking at the general attitude of the stratigraphic succession. Cross-bedded sets can also be bounded by parallel surfaces of higher rank; they are qualified as *master bedding* and constitute good datum planes. Remember, however, that master bedding is not horizontal in all cases (see clinoforms).

(G. Piacentini, 1970.)

Plate 36
Herringbone cross-bedding

This is a specific type of cross-bedding, which can be regarded as an *indicator* of process. In the vertical sequence of laminasets, you can note that two opposite dips alternate. This reflects a 180° change in the current direction, which is typical, although not exclusive, of tidal currents. Flood and ebb currents have opposite directions and also, in this case, the same power of transporting sand and making dunes. Each of them bevels the top of the previous bed, producing an erosional truncation of foreset laminae followed by the deposition of a new sand bed. From time to time, discontinuous mud drapes remain preserved in the sand (see color plate 18). Tidal currents mechanically predominate over waves in environments like estuaries and tidal flats, or in more limited areas such as the tidal inlets separating barrier islands within lagoon-beach complexes.

Not always do tidal currents produce herringbone bed-

ding. It frequently happens that one of the flow is much more powerful than the other, which is not able to carry a sufficient amount of sand to build up its own bed forms (see plate 34 **B**). The subordinate flow can, at the most, modify the structures produced by the main one; for example, it partly erodes them forming the already quoted *reactivation surfaces* (plates 26 **B, C;** plate 33).

In some cases, opposite tidal currents have the same power but spend it along different paths: some channels, for instance, are used by the flood current, others by the ebb current. No herringbone is then found, and one should look for other indicators of tidal setting.

In this picture, the laminae do not show the true dip angle because the section is oblique to the flow direction.

Pliocene intra-apenninic sandstones, Val Marecchia near Talamello, northern Apennines.   Photo: G. Piacentini 1970.

Plate 37
Bed forms: antidunes

This structure is still mysterious as geologists do not know exactly how it fossilizes, or whether it fossilizes at all. By its nature, it is an ephemeral configuration of a sandy bottom subject to current shear. In contrast to ripples and dunes, antidunes are not observable in static but only in dynamic conditions, as exemplified by the picture. They look like broad undulations with little relief, which appear on a flat bed, stay or migrate for a while, then

disappear and reappear again. In a sense, antidunes are pulsating forms, and can be produced in flumes, under controlled conditions; their formation depends on a certain combination of flow velocity and water depth. Actually, if one keeps the depth constant and increases the velocity of the current, the following sequence of events can be observed: plane sand bed, formation of ripples, formation of dunes, dune cancellation, plane bed with a

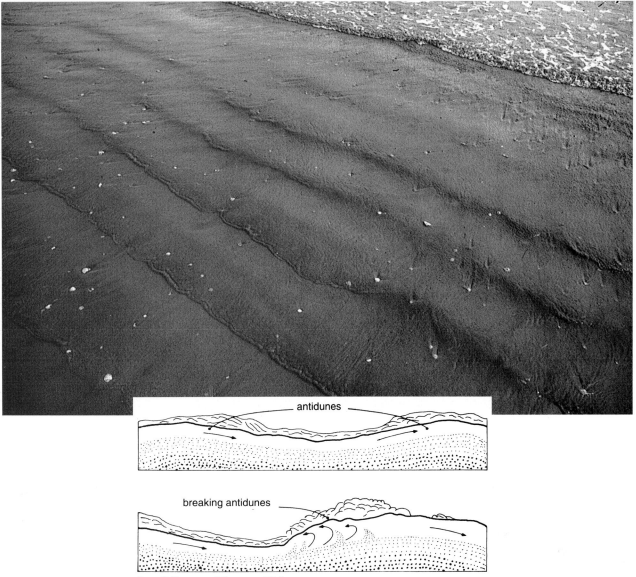

antidunes

breaking antidunes

(from Collinson and Thompson 1984)

carpet of moving grains, formation of antidunes, cancellation of antidunes and "pulsations" of the bed surface. For a given velocity, on the other hand, antidunes can be obtained by decreasing the water depth: as the picture shows (it was taken on a beach, when backwash was active), antidunes form under a thin veneer of water. The incoming wave thins out when climbing up the beach slope (swash), then gravity produces a return flow (backwash); in practice, we get here an oscillating current. The right combination of water velocity and depth varies within a certain range, which is known by hydraulic engineers as *upper flow regime*. A plane bed with moving grains is included in it, too. A static plane bed, ripples and subaqueous dunes are, instead, the expression of a *lower flow regime*. Upper and lower can be replaced by super critical and sub critical, respectively.

Besides the relief, what distinguishes antidunes from dunes is the absence of a well-pronounced slipface and of a shadow zone (boundary flow lines keep contact with the bottom), and the fact that they can be either stationary (standing antidunes) or migratory forms. When moving, they migrate upcurrent, which is the reason for their name. This means that sand is taken from the downcurrent side and deposited on the upcurrent side: in the picture, the water is flowing back to the sea, while the antidunes are moving landward.

We do not know of certain, well-documented examples of *fossil* antidunes: the profile of these bed forms is not preserved or, if it is, the large ratio between length and height (giving too small a relief or too great a length: the height of the examples shown here is of a few millimeters only) prevents its recognition in normal outcrop conditions. In section view, a partial preservation of antidunes is expected to show slightly inclined laminae. Low-angle lamination, however, is produced by other forms as well (hummocks, bars, etc.), and may not be considered as indicative of antidunes only. Moreover, the preservation of laminae itself is in question, because the periodic smoothing of the bed forms, preceded by breaking of water waves above them, remobilizes the sand with violence and puts it into suspension.

Inset: modified from Collinson and Thompson 1982.

A

B

Plate 38
Dunelike bed forms (dunoids)

These are dunelike structures that are typical of pyroclastic deposits and occur either as isolated or serial forms. Contrary to antidunes, they fossilize quite easily, but are never seen on exposed surfaces as fine ash buries them immediately after their formation. This obviously reflects their origin from catastrophic flows that carry a great amount of particles in suspension and rapidly dissipate their energy. Dunoids (a term proposed here) can thus be examined only in section view, where they appear as sets of inclined laminae with a wavy surface on top, draped by an ash bed (see also plate 39, and color photos 4, 15).

Why not call these structures dunes or sediment waves? The latter term is discouraged because of the scale, which should be much larger. Dunoids are more similar to dunes in both scale and shape, with a difference that consists in the lower inclination of the slipface: the dip angle rarely exceeds 15°, being thus smaller than the critical slope. Furthermore, the profile is smoother, with the foreset laminae joining tangentially both the trough and the crest areas with a sygmoidal trend (138 **A**). These characters suggest that skin-friction lines did not separate from the bottom, preventing the formation of shadow zones, which are typical of dunes and ripples.

The laminaset forming the body of the dunoid is not frequently truncated (138 **B** and plate 39, inset); when this happens, however, erosion can be more intense on the back than on the top of the form, causing an inversion of its profile. The upcurrent side becomes the steeper one (and, in this case, is erosional, not depositional), which make the structure similar to so-called chute-and-pools obtained in hydraulic experiments.

Tractive structures do not occur in all pyroclastic deposits, but characterize those which result from flows of high velocity and turbulence, where particles are not too concentrated and can be sorted according to size and weight. Coarser grains (pumice, lapilli) are concentrated near the bottom, and their burial by finer particles is delayed by the strong turbulence, which keeps them suspended for a time sufficient to develop bed forms. The flows originate from the base of ascending columns of gas and ash, and are energized by both gravity and the push of volcanic explosion; they are named *base surge* or *ground surge* (often but improperly abbreviated as *surge*). Other flows produced by explosive volcanic activity, though carrying particulate material, move more in the fashion of lava flows. Turbulence is here damped by the higher concentration of particles and the viscosity of the medium (a mixture of hot gas with fine ash), and little

or no vertical sorting occurs. The particles are deposited and buried quickly, or "freeze": no traction is possible, no tractive laminae are produced. This other mode of transport has been termed *pyroclastic flow,* and is mechanically similar to a debris flow (see plates 69, 74); the thermal state (presence of hot gas and lava particles, and water as vapor) makes the difference. In specialized literature, the attribute "pyroclastic" is frequently dropped, and only *"flow"* remains; this may be convenient in writing, but is a bit nonsensical and confusing. Flow is a generic term employed for every moving fluid substance; it cannot be appropriated by the jargon of any specialized area.

The lamination in pyroclastic deposits, when present, is particularly well defined, as shown by the pictures in this and the following plate; this is less the merit of the picturegraph than of the structure itself. The definition of laminae depends on the efficiency of traction, and this efficiency increases with a decreasing density of the medium. Such is the case for volcanic gas and vapor in comparison with normal air (see eolian deposits) and especially with water. A thinner fluid means less buffering of collisional effects between grains, which leads to better sorting and more effective abrasion. The separation and distinction of laminae are function, as I have already said, of the degree of sorting that occurs when grains are dragged along the bottom and collide many times.

The outcrops are in Salina, one of the Eolian Islands in the Tyrrhenian Sea. They still lay on the flanks of the original volcanoes, active in the Pleistocene.

Plate 39
Cross-bedding in pyroclastic deposits

A quarry wall in Lipari (Eolian Archipelago) shows the bedding style of a thick accumulation of pumice fragments (pebble-size particles of whitish, light and very porous volcanic glass) interbedded with some finer ash layers. The ash drapes some bed forms, which are cut oblique to the main flow. Two dunoids in a row are visible at half height, with internal cross-laminae following a sygmoidal pattern. Other isolated forms, above and below, are back-scoured and resemble chute-and-pools (see also inset).

Scale is absent in the picture: the approximate thickness of lensoid beds is 20 to 40 cm., as inferred from more accessible outcrops nearby.

Plate 40
Planar, low angle lamination

This small trench, about 35 cm deep, was dug across a beach crest (berm), separating the subaerial beach from the swash zone (to the right). Heavy minerals, rich in metals, were concentrated by the wave sieving into dark laminae; for this reason, lamination is particularly well defined, and records the vertical accretion of the sandy beach. The metallic minerals can reach economic concentrations and form sedimentary ore bodies (placers).

Most laminae form a plane-parallel, horizontally laying tabular set. In the upper portion, on the other hand, inclined surfaces and angular contacts can be seen. They indicate the presence of erosional truncations, caused by occasional or seasonal storms. During storms, the water level rises and the mark line shifts landward; at the same time, water agitation due to breaking waves increases. Sand is removed both as bed and suspension load, and is carried seaward. With the return of fair weather, the berm and the swash zone reoccupy a lower, seaward position and the sand is slowly brought back to shore along the sea bottom. Erosional surfaces are then covered by new sets of laminae.

Overall, an accreting beach will show in this part (foreshore zone) depositional features (laminasets) alternating with erosional features (truncations). Both depositional and erosional surfaces are planar and make low angles with the horizontal; they thus outline a stratification style made of intersecting wedges, which can be defined as *wedge-shaped bedding,* and is associated with *planar, low-angle lamination.*

Swash zone laminae can dip either landward or seaward, depending on whether they were deposited on one flank or the other of the littoral berm. Dip angles are, in any case, well below the critical angle of repose. The interface is, in fact, almost continuously swept by waves, which keep on it a super critical flow regime owing to the thinning water cover; on the backward side of the berm, however, wave energy is rapidly dissipated as water infiltrates through the sand, and a relatively steeper angle can be reached. Coarse sediment particles, shells, flotsam and jetsam are abandoned there, especially in storm berms.

At the toe of the swash zone, where the incoming waves and the backwash collide and water depth increases, the hydraulic regime becomes sub critical and ripples appear on the sand (plate 31).

Photo: A. Bosellini 1992.

Plate 41
Plane-parallel laminae in littoral sands

A belt of littoral sands of Pleistocene age ("yellow sands") borders the foothills of the northern and central Apennines along the Adriatic (northeastern) side of the chain. The sand is fine grained, well sorted, fossiliferous and locally contains pebbles. It has a poor degree of cementation, except for some levels parallel to bedding and laterally discontinuous (as shown by the picture).

Bedding planes are apparently parallel, but at a closer inspection they appear to converge and diverge at small angles. Stratification is actually wedge-shaped, and all beds are laminated. Skeletal remains are strewn on the planes of laminae. We can infer that this is the section of a fossil beach, whose facies and growth model are similar to those of the Modern foreshore shown in the previous plate.

Plate 42
Plane-parallel laminae in turbidite sandstone

The accreting surface is horizontal in this case, too, and the laminae very similar to the ones shown in plate 40. The environment of deposition was, however, completely different: a deep sea bottom, normally covered by mud but occasionally invaded by sand carried by a turbidity current. This shows us that planar lamination indicates a certain hydraulic condition (tractive flow in upper flow regime, phase of plane bed), which can occur in different environments of deposition. Therefore, it is an *indicator* of process, or mechanism, not of environment.

You may have remarked that plane-parallel lamination is not the only structure displayed by this picture. Other features are: 1) a vertical size grading in the lower part of the bed; 2) the presence, at the same level, of aligned dark fragments of pebble size: they are chips of clay or mudstone; 3) some sets of laminae (at least three) show undulations and small cusps: these are deformations, probably induced by water escaping during deposition.

Although not dominant and eye-catching, the above mentioned characters are significant: they give us more information concerning the process. The grading and the deformed laminae, for example, indicate relatively rapid deposition, which rules out a "particle by particle" accretion; the clay chips reveal that, not far from here, a muddy bottom was eroded by the agent responsible for deposition. All this makes us think of a possible catastrophic process; not one of too high particle concentration, anyway, because the well-developed lamination clearly indicates traction, and we know that traction is ineffective if sedimentation is too quick. Without other independent evidence, however, we can just suspect, but not prove, that this is a turbidite. Catastrophic processes, like storm waves and currents, operate also in littoral and shelf environments.

The sandstone outcrop is part of the Franciscan Complex along the California coast.

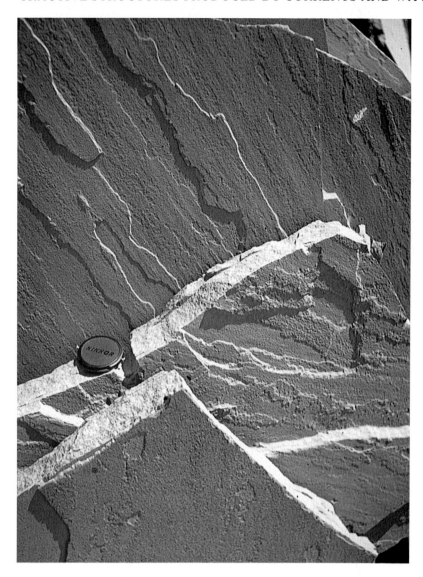

Plate 43
Current lineation on parting planes

Laminated sandstones can split apart along surfaces roughly corresponding to planes of laminae. Primary lamination and the ensuing diagenesis impart *fissility* to the rock, a property that helps both natural weathering (for instance, through freezing and thawing) and artificial sawing in quarries. Parting planes are smooth or slightly terraced. In both cases, they show, when the incident light makes a small angle with them, parallel strings of sand grains. This is a primary character produced by the tractive flow, which sorts and aligns the grains according to their texture and hydraulic behavior. A minor structure results, which is parallel to the paleoflow and is called *primary lineation* (to distinguish it from other kinds of lineations caused by tectonic processes). Which way the fluid moved along these lines cannot be detected.

The "fresh" surface of the sandstone slabs in the picture (no lichens, no coating of altered minerals) makes clear that quarries are privileged places for looking at rocks. They provide artificial outcrops, often of large size, in areas where the rocks are not or less easily observable because of vegetation or soil cover. On the other hand, quarries are dangerous places, whether they are active or abandoned. Visits must be authorized, and caution is needed in walking too close to cliffs and tailing piles: there is a certain risk of block fall or slide.

Quite often you are not welcome in quarries not so much for security reasons but because owners and managers suspect you are an ecologist or some kind of officer trying to discover infringements to law or environmental wrongdoing.

Plate 44
Vegetal debris on parting planes

Particulate vegetal matter of variable size is abundant in clastic sediments accumulating near swamps and marshes, for example in alluvial and deltaic areas. Owing to their low weight, even large fragments (dead twigs, etc.) are transported in suspension or float in the currents. Their deposition is delayed until they become water-logged, and their sizes are larger than those of detrital grains with which they are associated (the *hydraulic* size is more important than the geometric size: particles of different geometry and weight can have the same settling velocity). Plant remains can be resedimented from basin margins into deep water, as in the samples illustrated here, which derive from turbidite beds.

Vegetal debris is recognizable in Ancient, compacted sediments for its brown or black color resulting from carbonization (coalification) processes. Sulfur originally present in the organic matter is often found as iron sulfide, which, upon oxidative weathering, is altered into yellow-red oxides. The two specimens shown in the picture have different concentrations of vegetal particles spread on lamination planes. A preferred orientation, with long axes tending to be parallel to paleocurrent, is observable in the sample to the right.

In section view, laminae rich in vegetal matter are recognizable for their dark color. If carbonized debris is abundant in entire laminasets or intervals of sandstone beds, it can hamper cementation and make the rock more friable and erodible there. The resulting re-entrants in outcrops could thus be mistaken for clayey interbeds.

Marnoso-arenacea Formation, northern Apennines

Photo: G. Piacentini 1970.

Plate 45
Bed forms produced by waves: oscillation ripples

Plate 45 **A** shows the top of a thin sandstone bed in Pleistocene littoral deposits of the Marche region, Italy. The markedly linear and parallel ripples have small length and symmetrical profile. Not only have the two sides of a ripple the same inclination (see also B), but this is smaller than the angle of repose of the sand. This means that the ripples were produced by a short-period oscillatory motion, exposing alternately the two sides to fluid stress.

A modest amount of energy is sufficient to form small ripples, which implies that they record the passage of gentle waves in very shallow water or of stronger waves at a greater depth. In section view **B,** laminae can show both symmetrical (hut-roof) and asymmetrical (one-way dip) arrangements. They are said to be form-concordant or form-discordant, respectively, in relation with the external profile. To be precise, the term *symmetrical ripple* should be applied to the external shape only, and the structure as a whole should be called *oscillation ripple* or *wave ripple* (as opposed to *current ripple*). One more reason for this distinction stems from the fact that wave ripples can have an asymmetrical profile. This happens when there is a difference in the power of to and fro motions a wave orbit. In their seaward movement, water particles are closer to the bottom and lose more energy by friction; after accomplishing an orbit, they do not reach their starting point. In other words, the wave orbits are not closed and a net transfer of water mass occurs landward; correspondingly, there is a net transport of sand in the same direction. It is as if the sediment were subject to a current, albeit discontinuous and incremental, and the ripples reflect it by developing a slipface on their landward side.

The sandstone of picture **B** is a peculiar one; it is made of fragments of volcanic glass reworked by the sea.

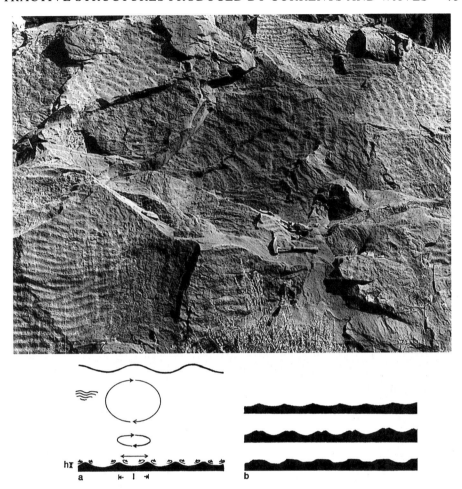

Waves attacked a lava flow emplaced under sea level, which was fragmented by rapid quenching and transformed in a detrital mass, a hyaloclastite.

Pleistocene littoral deposits, Lipari island, Tyrrhenian Sea.

Wave ripples are *environmental indicators*, at least in part, i.e., concerning water depth.

We find them under a shallow-water cover (a few decimeters to meters), in littoral, lacustrine and fluvial environments. Waves are not able to form ripples below a critical depth, the so-called *wave base*. This depth is related to wave length and power. Longer waves have a deeper base. Waves are produced by the wind, which transmits energy to the water by fluid drag exerted at the air-water interface. Small waves generated by a breeze on a small, ephemeral pond, are sufficient to create small ripples.

Plate 46
Wave ripples on vertical beds

The upper surfaces of several sandstone beds allow us to observe some variety of ripple forms and sizes. Both symmetrical and asymmetrical ripples are present on different bedding planes, the majority of them belonging to the oscillatory type produced by wave action.

Drawing **a** shows how the orbital motion of a wave transfers a shear stress to the bottom. The orbit flattens downwards until it becomes a to-and-fro linear movement.

The silhouettes of symmetrical ripples (**a** and **b,** top) tend to a trochoidal curve in equilibrium conditions. This shape is not symmetrical with respect to the horizontal plane, whereby it can be used as a way-up criterion in tilted beds (crests have a smaller curvature than troughs). When ripples are modified by water shallowing (see plates 31, 51–53), the polarity is recognizable as well.

direction of sole marks

erosional base of
beds and laminasets

Inset: Modified from Collinson and Thompson 1982

Plate 47
Hummocky cross-bedding

This term defines a variety of low-angle cross-lamination in which laminae are not planar but curved surfaces. Both convex and concave curvatures are recognizable: part of them is outlined by depositional laminae, and part by erosional surfaces. The preservation of these bed forms is thus partial, because some cancellation is inherent to their formative process. The cancellation is normally selective, affecting the protruding parts (hummocks) more, but does not spare the intervening depressions (swales). Hummocks and swales are not linear features like ripples and dunes (see inset). They have not been seen or photographed on the sea bottom, and their geometry has been inferred from stratigraphic sections. Cutting the structure at various angles, for example, one never gets a preferential dip direction of laminae; they have a quaquaversal, or periclinal attitude, and the same low angle of inclination in every direction. These forms, then, are not migratory, or little so. In scale, they are mostly in the range of dunes (see an exception in next plate).

Hummocky cross-bedding is regarded as the product of strong waves impinging on the bottom during storms. These waves have often the power not only of canceling ripples and putting sand in suspension, but also of eroding the bottom. Storm waves are highly turbulent, which means that sand can remain suspended while conditions of maximum energy persist (normal waves support only mud). When water motion and turbulence decrease, sand grains settle but are entrained horizontally by still strong oscillatory movements. Tractive laminae thus develop, although traction cannot last long as the energy wanes and more sedimentation takes place. The formation of hummocks should be, in conclusion, a short-lived process associated with a catastrophic event. The peculiar dynamics of this kind of events explain why the hummocks do not remain visible on the sedimentary interface: they are either buried by the finer suspended sediment or remolded by normal waves (overprinting them with ripples), or both.

When hummocks are eroded by subsequent erosion, concave-up surfaces are preferentially preserved: the structure is then described by the term *swaley cross-bedding* (or flat festoons).

The outcrop is part of the Pleistocene littoral sands on the Adriatic margin of northern Apennines

Plate 48
Small-scale hummocky cross-lamination

Sets of low-angle convex laminae are here truncated by slightly curved surfaces. The thickest laminaset (below the pipe) shows a preferential dip to the left, which suggests a certain amount of lateral accretion of a hummocky bed form. In cases like this, one can invoke a unidirectional component of movement, i.e., a current. Actually, storms are stimulated by strong winds, and these winds can produce not only waves on the water surface, but also currents down to a certain depth. Such currents sometimes have a tractive capacity; that is why many sedimentologists think that hummocky cross-bedding derives from a combination of wave and current action.

This is the same outcrop as in plate 45 **B** (Quaternary volcaniclastics reworked in the littoral zone); we see, then, that hummocky forms and normal ripples are associated in this unit. Hummocks formed during storm conditions, ripples during fair weather, when normal waves reworked the storm deposits.

Having previously defined the concept of wave base, I must now specify that there are at least two such levels: one for normal waves and one for storm waves, the latter being obviously deeper. Whereas ripples can form only above the fair weather wave base, hummocks form both above and below it. Shallower hummocks are canceled, but deeper ones escape destruction and are preserved under mud drapes.

Other examples of small-scale hummocky cross-lamination are illustrated in color photo 5, from Pliocene sandstones of northern Apennines. Storm beds are found in submarine beaches exposed to the open sea, in shelves, and in protected water bodies (bays, lagoons, large lakes). When storms invade tranquil environments, their deposits and related structures are more easily preserved thanks to the absence of mechanical reworking during normal conditions.

Hummocky and swaley cross-bedding, in conclusion, are structures indicative of traction and deposition under storm conditions. As such, their presence indicates a *storm layer*, or *tempestite*. Not all tempestite beds, however, contain hummocky or swaley cross-stratification. Pay attention to this type of logical connection, which is frequent in geology: feature A implies process B, but B implies either A or other characters, depending on circumstances. In other terms, there is no biunivocal link between cause and effect. This must be clear for a correct understanding and use of the features that are qualified as *indicators*.

Plate 49
Hummocky cross-bedding: rhythmic pattern

The vertical alternation of less and more cemented beds emphasizes the rhythmic stratification style and its concave-convex geometry in this outcrop of Pliocene sandstones in the northern Apennines. The difference in cementation is a diagenetic character but reflects primary textural variations within the sediment. The sand is slightly coarser and cleaner in cemented beds: more pore space was available for chemical cement in it. In softer beds, on the other hand, some *matrix* (collective term for fine particles) is admixed with the sand and occludes part of the pores: less cement was precipitated there, also because the sediment was less permeable to percolating water.

Cemented beds have also a sharply defined base, whereas their top grades into matrix-rich sand. One can assume, to explain these features, that storm waves scoured a sandy bottom removing both sand and finer particles. Relatively sorted sand was redeposited first, while the fines were kept in suspension. With waning wave energy, fines were increasingly added to the sand and the upper member of the couple was accumulated, thus completing the storm layer. Hummocky and swaley laminae are preferentially preserved in cemented beds, but not visible from this distance (the cliff is an inaccessible river trench, incised during a rejuvenation phase of the Apennines chain). The presence of the structure, however, gives the bedding its peculiar configuration, with broadly curved, laterally discontinuous surfaces.

Pliocene Intra-apenninic Basin, Zena Valley, northern Apennines.

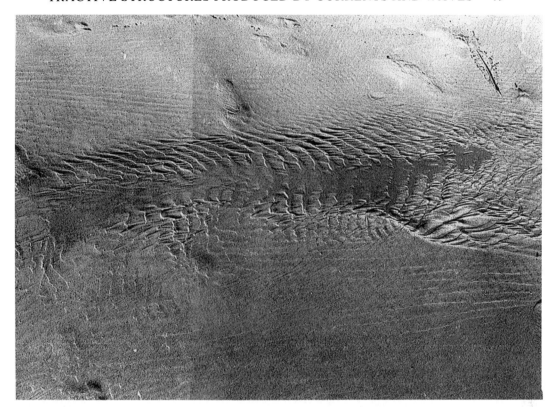

Plate 50
Modified tractive structures: lozenge-shaped ripples

Bed forms and tractive structures have a highly variable *preservation potential,* depending on both the dynamics of the environment and geologic factors such as subsidence, sea level changes, sedimentation rate, tectonic movements, etc. Once formed, they can be 1) canceled, 2) buried and preserved integrally, and 3) altered or modified, and preserved in this modified state.

Various processes can modify the structures: many of them are mechanical, but chemical and biological processes are also effective. Chemical and biological modifications will be examined in the sections dedicated to diagenetic and biogenic structures. *Modification structures* could constitute a separate item, because they are often very useful as environmental indicators. What they indicate is a change in the environment, occurring soon after their formation or after a time lag. A structure formed under shallow water, for example, can emerge if the water evaporates or is drained away. Some environmental changes are more drastic than others; moreover, they can involve the whole environment or some parameters only (e.g., the rate of sedimentation, the ventilation, etc.).

Let us see how beach structures are modified when the sea retreats at low tide. Accretionary beaches, where the supply of sand is abundant, are wide and have a low topographic gradient. Their submerged portion is often characterized by a wavy morphology, with bars or ridges forming the relieved portions. These long shore bars and ridges are parallel or oblique to the shoreline; the shallower ones make the incoming waves break and emerge at low tide. During shallowing and emergence, subaqueous ripples tend to be canceled or to have their crest smoothed, as the thinning of the water cover changes the hydraulic regime from sub critical to super critical. In depressed areas between the bars (troughs, runnels), some water can remain; ripples are not smoothed out completely, but modified or distorted, as in the example shown here.

A small and shallow channel drains water toward the sea (to the right). The ripples maintain a transversal orientation in the central area and are oblique near the sides; their relief is minimum. Water films coming from different directions interfere and tend to split the ripples into rhomboid segments. On the far right, linguoid and spatulate (spoon-shaped) forms develop.

A

Plate 51
Modified tractive structures: interference ripples

The summits of littoral bars gradually emerge at low tide and form small islands. The wave fronts, which are rectilinear and approximately parallel to the shore during average or high tide, are refracted by the emerging bars when the water shallows. They tend to become parallel to the island contours and, in so doing, split into overlapping and intersecting arcs.

The consequence is that the local direction of water motion changes with time, and the ripple crest orientation is forced to adapt to these changes. However, to cancel the ripples that are not in equilibrium with the new conditions, a certain mass of sand must be removed, while waves are losing energy. The "old" ripples are thus simply modified, and new ones are printed over them, with crests making an angle with the old ones. More generations of overprinted ripples can be formed, each one being smaller than, and intersecting the previous one.

The whole pattern receives the name of *interference ripples,* even though *intersecting ripples,* or *overprinted ripples,* would be more appropriate. The various wave

B

systems are, in fact, successive in time. Wave interference, anyway, does exist, as shown by the larger picture; it rather produces its own equilibrium forms, such as the rhomboidal ripples of the previous plate, which replace, instead of overprinting, the older ones.

On the basis of what has been said above, try to determine how many generations of ripples are present in picture **B,** and their order of formation.

Photo: Piacentini 1970.

Plate 52
Interference ripples

The use of the term ripple for this polygonal structure seems puzzling, ripples being essentially linear structures like those we have seen in previous images. This honeycomb pattern was once interpreted in another way, i.e., as a biomechanical structure named *tadpole nests*. Actually, tadpoles living in shallow ponds move by waving their tails. The slight agitation of the water is sufficient to move fine sand and silt grains below the small organisms and to accumulate it aside. The competition for space between individuals of about the same strength produces pits of about the same size, whose shape is dictated, as in the case of honeybees, by optimal utilization of space.

If one looks carefully at the picture, one can detect, throughout the polygonal pattern, certain alignments of crests suggesting that the structure could also represent the interference or intersection of small ripple systems with different orientation. In any case, this curious structure is valuable as environmental indicator: it formed under an extremely thin water cover, and the environment of origin was emergent or almost so (from littoral-intertidal to continental).

The specimen is a thin slab of siltstone from the Old Red Sandstone, U.K., and belongs to the collection of the Department of Geological Sciences, University of Bologna.

Plate 53
Modified bed forms: double crested ripples

Small waves in thin water can modify ripples in this way (see inset b, plate 46 for the corresponding profile). Apparently, the wave fronts did not change direction but simply tried to make more closely spaced ripples. They were successful only on crestal zones as water was too deep in troughs. No new orientations are superimposed to the old generation in this case. It is a form of low-energy modification.

Other forms of modification in coastal and nearshore environments are represented by ripples that rework the topmost part of storm deposits. An example is portrayed in color photo 17. Remember that small-scale structures induced by waves are found only above fair weather wave base, while medium to large-scale structures occur farther offshore up to storm wave base. In both cases, I am talking about average base levels: exceptional storms or earthquake-induced tsunami waves can stir water and sediment up to 200 m depth.

CHAPTER 3
Structures Produced by
Traction-Plus-Fallout

Plate 54
Climbing bed forms: ripple-drift lamination produced by currents

Bed load and suspended load can give both separate and combined deposits. Distinct beds often alternate in time on the same place, as we have already seen in many examples in the first section. On the other hand, traction and fallout can cooperate in accumulating a single, almost simultaneous deposit. When bottom traction of coarse material is accompanied by deposition from suspensions, tractive structures still form but acquire some peculiar features that allow us to recognize the role of the suspended load. They are called *traction-plus-fallout structures*.

Traction and fallout can act together but are also competitive mechanisms: the more particles settle, the more hampered are their movements on the bed. In other terms, with a few particles in suspension, one gets purely tractive structures; with a high particle concentration and a high rate of sedimentation, traction tends to be suffocated. The picture shows an intermediate condition, where tractive laminae are well developed and fallout subordinate. The bed is a turbidite sandstone, i.e., the product of a catastrophic flow, and it can be assumed that most of this sand was transported in a turbulent suspension: to keep sand suspended, a high turbulence is needed, which implies a high flow velocity.

The continuous addition of sand to the bed makes it accrete while ripples are migrating on it. How this sand is distributed determines the geometry of the resulting structure (lamination). If the sand is deposited on the lee side of ripples only, while the stoss side is kept free or eroded, every ripple advances by climbing on the back of the adjacent one. An incremental ramp thus forms, whose cumulative expression is a diagonal surface, dipping opposite to the flow. Such surfaces develop from each ripple crest; they are parallel and sometimes more pronounced than the foreset laminae, thus giving a false impression of inclined bedding. This effect is not apparent in this picture: surfaces of climbing, as they are called, are sharper in plate 8.

To the right of the pen, climbing ripples "degenerate" into oversteepened forms (see also plate 118); this means that sand grains locally stuck together and responded as a coherent mass to current drag.

Other examples of climbing ripple cross-lamination can be seen in color photos 7 and 16. In photo 7, the sand is very fine and mixed with silt, and dark hues in the lower foreset and bottomset laminae indicate concentrations of vegetal matter and clay.

Marnoso-arenacea Formation, northern Apennines.

Plate 55
Climbing bed forms: truncated and sinusoidal ripples

Let us now examine a situation where grain fallout takes the lead on tractive effects. With increasing rate of deposition, some sand and silt accumulate also on the stoss side of ripples, and backset laminae form (**A**). Their thickness increases as to equalize that of foreset laminae (**B**). Ripples keep on climbing, but the angle of climb becomes steeper. This means that the migration of bed forms decelerates, while the vertical accretion of the bed speeds up. The more balanced sediment deposition on the two sides of ripples reduces the asymmetry of their profile, which has been called *sinusoidal*.[1] Laminae associated with sinusoidal ripples are continuous and have a draping attitude; the term cross-lamination becomes improper at this stage, because there are no more erosional truncations. It might be called *asymmetrical wavy lamination*. On the other hand, this is the end member of a continuous spectrum of phenomena related to bed form migration and tractive lamination, which is centered on cross-lamination. Moreover, wavy lamination rarely occurs in separated beds; it usually grades into genuine cross-lamination, as these pictures show.

The outcrop is artificial and is found in a sand pit open in Pleistocene glacio-lacustrine deposits in Massachusetts. The sand was carried into lakes by short-headed streams fed by melt water.

Climbing bed forms originate mostly at small scale; examples of medium scale are found, although not frequently, among subaqueous dunes and pyroclastic dunelike forms (see color photo 15). Suitable conditions for generating larger forms are probably attained with more difficulty in comparison with ripple-generating suspensions. The scale, strength, and persistence of turbulent motion required for larger forms are rather exceptional (e.g., violent volcanic eruptions).

Plate 56
Climbing ripple and fallout lamination

Ripple climbing changes modality in this single layer. The ratio between traction and fallout starts from a maximum at the base, then decreases upward; the morphology of ripples and lamination vary accordingly. First, you see the variety with truncated stoss sides, occupying an interval of some centimeters. The laminae become then more continuous and drape over sinusoidal forms that are increasingly symmetrical. In the central part of the deposit, undulations die out and plane-parallel lamination gradually replaces and definitely buries the cross-laminated interval. At the same time, silt grade particles substitute for sand in the settling sediment. Not only is the current weaker when fine grains are deposited, but the sediment

is also sticky and resists traction as a coherent mass. That is why ripples cannot be formed any longer.

In the lower part of the layer, lateral transitions can also be observed between more to less tractive forms, and vice versa. This indicates local changes either in the current strength or in the fallout rate or in the physical properties of the newly deposited sediment (which has a critical texture, at the boundary between sand and silt).

Overall, this layer records the passage and deposition of a waning current, probably a flash flood in an ephemeral stream bed (ouadi). It occurs in Recent continental sediments of Oued Saùra, western Grand Erg Occidental, Algeria.   *Photo: G. G. Ori 1992.*

Plate 57
Traction-plus-fallout structures: vertical sequence in flood deposits

Waning flows carrying a great amount of suspended sediment are usually associated with catastrophic phenomena like gravity-driven turbidity currents, storm or tsunami waves at sea, fluvial floods. Tractive transport represents only a moment, a phase in the development of mechanical and hydraulic conditions promoted by such surgelike events. They strike a tranquil bottom with concentrated energy and pass by like a train, leaving a residual turbulence in their tail. During their passage, the local energy decreases, more or less rapidly, after the initial peak. A surgelike, catastrophic flow leaves a quite distinct signature in its deposit, especially if it is turbulent. An example is a turbidite layer, which we first met in plate 8. Another is presented here, and derives from a fluvial flood as the one in plate 56.

Flood waters made a breach in a levee of a channel; water passing through this breach, or crevasse, formed a secondary current that expanded in the adjacent alluvial plain. The flow expansion (jet effect) caused a slowing down of the current and a loss of transport capacity.

Sediment was thus abandoned in a deposit called *crevasse splay*. Deposition can be preceded in these cases (as in other cases of catastrophic flows) by erosion, which is reflected in a sharp base of the bed. Coarse sand and gravel can be strewn on this erosional surface, as a part of the bed load can escape from the main channel through the crevasse (when levees are overtopped, instead, only suspended load gets out). Finer sand and then mud can follow, thus forming a *graded bed*. The grading of the grain size and the sequence of traction-plus-fallout structures (analogous to the "Bouma sequence" of turbidites, see plate 8) constitute the signature of the waning current. Plane-parallel laminae are here on top of coarser textured, low-angle laminae, and are followed by ripple-drift cross-lamination. Above the white undulated surface, which is the top of the event, another bed starts directly with tractive structures (dune cross-bedding).

Gravel pit in Recent alluvial deposits of the Jarama river, near Madrid, Spain.

Plate 58
Fallout deposits: mud drape on ripples

A tractive current with little or no material in suspension will leave bed forms exposed on the top surface of sand. When the sea retreats at low tide, for example, ripples, dunes and other forms are visible, as documented by previous pictures. If mud is present, it will settle on the tractive deposit when the water becomes calmer. An abundant suspension load tends to bury the bed forms under a relatively thick, flat-topped layer; in section view, a gradual upward transition from sand to mud will be observed (see plate 56). When, on the other hand, mud is more diluted, its deposition is slower or delayed; the mud forms a thin drape under which the underlying morphology is recognizable, as in this picture showing a river bed after the retreat of flood waters (pencil for scale to the lower right).

On drying, the mud shrinks and cracks in a typical polygonal pattern (see also plates 104–106). With further loss of water, the edges of the polygons curl up (plate 105, color photo 27) until clay flakes detach from the sandy substratum (color photo 28). *Mud clasts* are thus prepared in place and can be carried away by wind or running water.

A couple of beds, made of sand and mud respectively, forms the basic element of rhythmic, repetitive successions, a sample of which is often called a *heterolithic facies*. The *pelitic* member of the couple is also referred to as interbed, interlayer, parting or drape. *Pelite* is a broad term encompassing all mud-size (silt to clay) sediments, regardless of their consolidation state: mud, mudstone, clay, claystone, shale, argillite. As for silt and siltstone, their coarser types are usually grouped with sand and sandstone, the finer ones with pelites.

Plate 59
Rhythmic stratification in fallout deposits

The bedding in this outcrop is enhanced by color contrasts superimposed on lithologic variations. In cases like this, the terms *band* for thin bed or lamina, and *banded deposits* in general, are often used. Strong differences in color or hue of sediments sometimes reflect only subtle changes in composition and texture: small amounts of pigments are, in many cases, sufficient to produce neat colors.

The bands in the picture vary in thickness from nearly 10 cm (white one at the base) to 1–2 mm. A question that can be asked is whether they represent depositional units of variable thickness but of the same rank (beds or lami-nae) or units organized in different hierarchical levels (laminae, beds, layers, bedsets). In other terms: do the thin bands represent small, independent events or momentary pulsations of a larger event? Answer is not easy in the present case. Some thick white bands have a sharp base and top, and should represent distinct beds (faint laminae inside confirm this impression). More problematic are the intermediate and thin bands. If you look at the lithology, three lithotypes are recognizable: 1) white, crystalline (saccharoidal) gypsum; 2) black shale rich in organic matter (kerogen, or bitumen), and sparse gypsum

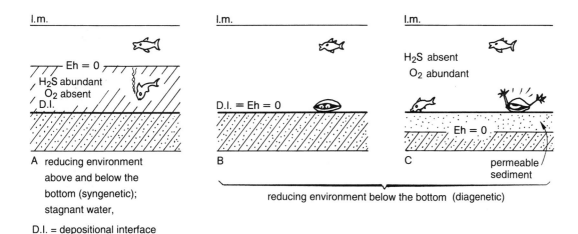

A   reducing environment above and below the bottom (syngenetic); stagnant water,

D.I. = depositional interface

B

C   permeable sediment

reducing environment below the bottom (diagenetic)

crystals not visible to the naked eye; 3) gypsum with clayey and organic impurities (various shades of gray).

The fallout of gypsum particles is part of the evaporitic process, starting with the nucleation of crystals at the air-water interface when concentration of salts in seawater is concentrated by strong evaporation. To complicate things, there is the possibility of the crystals (which form a sort of mush on the bottom) being reworked and redeposited. Gypsum beds could thus represent either (or both) primary precipitates or clastic deposits (gypsum arenites and gypsum pelites if sand or mud size particles prevail respectively). It can be hypothesized, for example, that the thinner bands, rhythmically alternating with the black shales, derive from direct evaporitic precipitation, probably seasonal (*varves*), whereas the thicker units are made of resedimented gypsum crystals and fragments. Fallout is anyway the dominant mechanism, because reworked particles apparently traveled in suspension.

In any case, gypsum deposition was not a continuous process but a periodic or sporadic one. It was hosted in a tranquil, protected environment, occupied by stagnant water. Conservation of organic matter tells us that oxygen was not present on the bottom (anoxic conditions), as confirmed also by the absence of organic activity. The lack of oxygen implies scarcity or absence of bottom ventilation and water circulation. Furthermore, the abundance of organic matter means that surface waters were affected by *eutrophication,* i.e., high production of biomass, mostly represented by phyto and zooplankton, favored by a high supply of chemical nutrients. An abundant production of biomass obviously means a mass mortality and a high sedimentation rate of organic remains.

Now, if one assumes that the nutrient input was associated with an influx of fresh water from rivers or of seawater of normal salinity, causing also a dilution of salts in the basin, it is easy to explain the alternation of conditions favorable to deposition of mud and organic matter on one side, evaporitic gypsum on the other.

Messinian evaporites (Gessoso-solfifera Formation) of northern Apennines, Perticara area.

The fallout mechanism does not produce particular structures, if bedding is excluded. In normal cases, it involves muddy sediment and tends to drape previous deposits. Fallout of mud can proceed at a slow pace, particle by particle, or be quickened by aggregation of individual particles and *flocculation.* Mud beds are texturally uniform or subtly graded. Porosity, and hence the water content in muds, is higher than in sands; initial porosity, however, is greatly reduced by burial and *compaction:* with increasing overburden, water is squeezed out of the sediment, whose density increases. Muds are commonly rich in clay minerals, which mostly occur as tiny, platy particles; other flat particles of larger size can be present, such as mica flakes and vegetal matter. All of them are originally deposited in a haphazard fashion (like a castle of cards), giving an isotropic fabric; during compaction, they tend to be aligned parallel to the bottom (i.e., normal to the gravity force). An oriented fabric develops, and the sediment acquires an aspect that can vary from foliaceous to scaly to laminated. Compacted, laminated muds become *shales,* or fissile mudstones. Fissility is a secondary property, related to diagenesis, but sometimes overprints a primary lamination (if organisms did not destroy it).

A *thick* fallout deposit with a homogeneous texture is not easily interpretable; it could derive either from a massive, instantaneous event, such as mud flow, or from a slow settling lasting for a long time in physically stable conditions, such as the descent of plankton remains in the quiet bottom of the deep ocean. Moreover, the structureless character of a thick pelitic bed is not necessarily related to deposition. It could be secondary, as mechanical (liquefaction) or biological disturbances might have canceled lamination or bedding.

Plate 60
Heterolithic bedding: wavy and lenticular types

Heterolithic practically means made of two or three alternating lithotypes, the most common being sand and mud (sandstone and mudstone). Bed thickness can vary; the examples presented in this and the following plate are thin-bedded varieties, in both Ancient and Recent sediments.

The sandstone beds, belonging to the Triassic Servino Formation of southern Alps, are laminated and ripple-molded at the top. Their bases can also be undulated by sinking of sand and compaction of mud in troughs of underlying rippled beds. The stratification style has been called *wavy bedding* by students of tidal sediments,[2]

where it is commonly, though not exclusively, found. Bedding is also said to have a *pinch-and-swell* structure.

We see also, in the picture, that thinner sandstone beds loose their individuality laterally by leaving place to an alignment of discontinuous lenses. This means that tractive energy was available, but sand was not, at least in amounts sufficient to cover the whole bottom. That is a second stratification style, called *lenticular bedding*. Ripples separated by mud stripes are *starved ripples*.

Wavy bedding is also shown in plate 61 and in color photo 6.

Plate 61
Heterolithic bedding: flaser and wavy types

A *flaser* is a thin lens of mud embedded in sand, with a concave-up curvature reflecting its being the remnant of a mud bed draping ripple troughs. The mud was eroded from ripple crests, which led to amalgamation of several sand beds (see introduction, figures 8C and 9); amalgamation can go undetected if sand composition and texture do not change from one bed to another, and flasers are not preserved.

In 61 **A**, *flaser bedding* predominates in the central sand body, passing to lenticular bedding both upwards and downwards. In **B**, you see an alternation of all three types of heterolithic bedding. Both photographs were taken in trenches dug in North Sea tidal flats.

Larger-scale flasers are found on dune troughs like those shown in plates 28 and 29. Fossil examples of small- and medium-scale flaser bedding are shown in

A

B

color photo 18, in an outcrop of a classical European formation, the Swiss Molasse near Fribourg. This sandy unit was deposited in shallow water influenced by tidal motion, on the bottom of a Miocene seaway bordering the newly emerged Alps chain. The depositional basin was an asymmetrically subsiding, elongate trough, called a foreland basin, which received its clastic supply from erosion of the Alps. The Alpine chain was built up, from the Cretaceous to the Oligocene, by the collision of Europe and Africa, which destroyed the interposed oceanic crust of the western Tethys and deformed the margins of the two continents, including many sedimentary successions.   *Photos: H. E. Reineck 1970.*

ENDNOTES

1. A. V. Jopling and R. G. Walker. 1968. Morphology and origin of ripple-drift cross-lamination, with examples from Pleistocene of Massachusetts. *J. Sed. Pet.* 38: 971–984.

2. H.-E. Reineck and I. B. Singh. 1973. *Sedimentary Environments.* New York: Springer-Verlag.

CHAPTER 4
Textural Indicators of
Transport and Deposition

Plate 62
Organized conglomerate

Rudites, including gravel (loose) and conglomerate (cemented), are the coarsest sedimentary deposits. They are frequently structureless, but even when they are present, bedding and structures are recognizable with more difficulty than in sands and sandstones. Bedding surfaces are often crude, rougher, and ill-defined; they are related to vertical changes in one or more textural parameters, such as grain size (average, maximum, range, and variability), shape and roundness, packing (closeness or spatial density), orientation and fabric, and type and amount of *matrix* (finer materials occupying spaces between pebbles).

In the picture, bedding is emphasized mostly by marked differences in grain size of pebbles. The pebbles are closely packed, which means that there is little room available for matrix or cement between them. A certain degree of cementation, however, occurred; it allows the maintenance of a vertical cliff, and the description of the sediment as a conglomerate.

The beds have a certain lateral continuity, at least within the limits of the picture frame; an exception is made by two small lenticular units near the top. The lenses have a concave base and a flat top, indicating that they represent the fill of local scours. Such lenticular beds of very limited lateral extent are also called *pockets*.

The picture is a close-up view of the clinostratified body illustrated in plate 14 (gravel pit in Pleistocene littoral deposits of Marche Region, Italy).

In general, rudites are qualified as *organized* when a certain order, structure or organization of constituent particles can be recognized in them, *disorganized* when they look like randomly heaped stones.[1] As said before, organization is shown by textural characters and fabric, for example sorting by size or shape, alignments or clustering of pebbles (see plates 66, 67), or particular orientations of pebbles (plates 64, 65).

Plate 63
Disorganized conglomerate

No bedding or particular fabric is recognizable in this outcrop of a Pliocene conglomerate in fluvio-deltaic deposits of the Pliocene Intra-apenninic Basin. Again, as in the previous example, pebbles are closely packed and touch each other; the particles are said to be *self-supported,* and the conglomerate *grain-supported*. With decreasing concentration (see, for instance, plates 66 and 69), pebbles become *matrix-supported,* or matrix-embedded. A matrix-supported conglomerate is also called a *paraconglomerate*.

Another observation is that size sorting is not bad but poorer than in preceding and following images (sorting is one of the parameters that define organization). In terms of shape, there is also a moderate sorting, with a predominance of globular, ellipsoidal forms and a small number of elongated and flat clasts. The lack of any fabric, i.e.,

the arrangement of pebbles in a quite haphazard fashion, is anyway the most impressive character of this conglomerate, which induces us to use the attribute disorganized (or poorly organized). The most obvious reason for it is a catastrophic emplacement by a mass flow, such as a fluvial flood of exceptional violence or a gravity flow. A more selective process would have produced some bedding or preferred orientation of pebbles. The flow energy must have been very high to carry the pebbles with so close a spacing and so little matrix (mud in the matrix lubricates the movement, and solid friction between pebbles is high).

The moderate sorting seems to indicate that the mass flow picked up materials that were pre-sorted in some local repository.

Plate 64
Pebble imbrication

When a high turbulence is produced by currents or waves, pebbles can be removed and kept in temporary, or *intermittent suspension*. Their weight does not allow them to travel long distances in that way; the pebbles rather make long jumps. Support is provided by fluid friction and by an arrangement similar to that of an airplane wing, i.e., dipping upcurrent. This orientation is maintained when the pebble lands and joins other companions in a shingled arrangement (*imbrication*). Imbrication is commonly observed in river beds, but can be produced also by the oscillatory motion of waves on pebbly beaches. In this case, the asymmetry of orbital motion, with its predominance of landward push, make the pebbles dip seaward.

Imbrication is best displayed by platy and discoidal particles, whereby a well-imbricated gravel is rich in such elements, as the picture shows. This can be an original character, when pebbles derive from fissile rocks, but is usually acquired by hydraulic selection (shape sorting) during transport. More commonly, in fact, the original population of pebbles is a mixture of various rock types, with different sizes and shapes. The dimensions of pebbles are defined by length (a), width (b) and thickness (c), which are measured along three mutually perpendicular lines. These lines are, in their turn, orthogonal to three planes cutting the pebble along sections of maximum, intermediate, and minimum area. To get these sections,

you do not need to saw the pebble: it is sufficient to project its contours on a plane from three view points at angles of 90°. Ideally, the pebble is equated to an ellipsoid, whose axes represent the three dimensions.

Platy or discoidal particles are those in which the "equatorial," or maximum area section (containing a and b axes) reaches its maximum development. The external surface intersected by c axis (normal to ab) is thus the one with maximum area, where most friction is exerted by both the fluid medium and other solids. It is the surface on which the pebble rests in its more stable position. On a smooth bottom, the pebble would lie with the plane ab horizontal; on a rough bottom, covered for example by other pebbles, the particles adhere to one another with dipping ab planes. This maximizes solid friction and resistance to being removed by a current. Imbrication is, in conclusion, a fabric of maximum stability; to dislodge imbricated pebbles, strong local eddies, sort of "hammer blows," are needed.

The pebbles seen in the picture were accumulated by waves on a beach foreshore (top of the outcrop in plate 14). Imbrication, plus size and shape sorting, make this an example of organized conglomerate.

Gravel pit in Pleistocene littoral deposits of Marche Region, Italy.

Plate 65
Edgewise pebbles

Plate 65 shows the emerged part (foreshore and back-shore) of a gravelly beach forming a narrow strip at the foot of a coastal cliff (Adriatic coast of Italy). The different effects of fair weather and storm waves can here be appreciated. The former ones sort the materials arriving to the beach from stream mouths or cliff erosion: they separate sand from gravel, disposing them in different parts of the beach profile, and remove finer particles by taking them away in suspension. Here we see small waves lapping on fine sand with scattered pebbles, then, proceeding landward, a string of coarse sand and small pebbles making an abrupt transition to larger pebbles.

Steeper waves raised by storms erode the sediment from the submerged portion of the beach, and bulldoze it, without much selection, on the emerged portion (which is then inundated), where they loose their energy. The sea retreats from the backshore after the storm, and normal waves do not modify the storm deposits (but walking creatures can). Arcuate ridges, representing storm berms, are thus preserved. In them, pebbles are stacked or wedged with steeply inclined or subvertical long axes. A prevalent seaward dipping can be observed in the foreground.

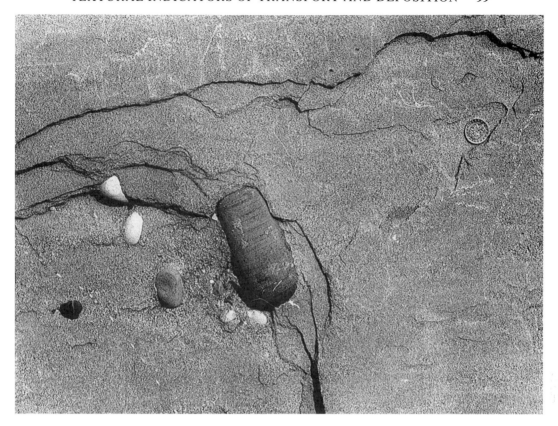

Plate 66
Pebble cluster

When a coarse particle remains anchored to the bottom and the current is not able to remove it, a shadow zone is created by its protrusion: boundary flow lines detach from the bottom at the summit of the obstacle, jump into the fluid as from a diving-board, and reattach farther on. Smaller particles can find a protection in the shadow zone, where turbulence and velocity are reduced, and come to a rest. In some cases, the jump (flow separation) occurs before the obstacle because it offers a sudden and strong resistance to flow (a wall effect), and a shadow zone is localized there, too. Depending on the size and the shape of the obstacle, various cases can occur: no shadow zones (or too small ones), presence of such a zone either before or behind the obstacle, or in both positions (compare with plate 82).

Pebble clusters are the signature of hydraulic shadow zones, even if the obstacles are later removed. The cluster seen in the picture is particularly evident because the pebbles are embedded in sand. When the whole bed is made of pebbles, one can recognize the clusters for their contrast in particle size or packing with the surrounding areas.

A further observation is possible here: it gives us a hint to the mechanisms operating in the bed load of a tractive current. The larger pebble has an elongate shape (prevalence of *a* axis over the other two), and is transversal to the paleoflow direction. This means that it rolled on the bottom. When similar particles are suspended, their long axes are aligned parallel to the flow.

Messinian sandstones and conglomerates of the Colombacci Formation, Pietrarubbia, northern Apennines.

Plate 67
Tractive shell beds

The sea has a tremendous amount of energy, part of which is spent in transporting sediment. It can happen that waves and currents are powerful enough to entrain coarse particles, but these are not available. Pebbles can be brought to the sea by rivers and glaciers, but remain rather localized near river mouths and other point sources. Far from them, the hard parts (shells, skeletons) of marine benthic organisms can play the hydraulic role of pebbles. They are removed by erosion, especially during storms, and redeposited in the same or a different environment. Organic, or skeletal rudites can thus accumulate; they have received a variety of names in geological literature, including shell beds and coquina beds.

The most violent storm events transport and deposit shells and their fragments in mass. Evidence of mass transport is given by the structureless and disorganized character of the deposit: lamination is lacking, particles are randomly oriented and packed, etc. Disorganized coarse sediments are also qualified as *chaotic*.

The shell beds shown in the picture are not massive but tractive, as indicated by two features: 1) the crude layering emphasized by the alignments of shells, and 2) their convex-up orientation.

The shells are mostly disarticulated valves of bivalve mollusks. Were they brought into suspension by a massive flow, either turbulent or laminar, a haphazard arrangement would appear upon deposition (or "freezing"). Under the effect of tractive motion, on the other hand, they assume a stable position on the bottom, as to offer a

minimum resistance to fluid stress; this implies a consistent orientation, statistically speaking, and that is what is seen in this outcrop. The shells are not only streamlined but well aligned in thin beds or pavements suggesting wave rather than current action. Unidirectional currents tend, in fact, to leave coarser particles more scattered between the finer ones.

Pliocene Red Craig Formation, Norwich, East Anglia, U.K.

Displaced organic remains are said to be reworked; when they are redeposited in a sediment of about the same age as the original one, geologists speak of *intraformational reworking*. *In situ* reworking can also occur, when, for example, normal waves erode, wash and sieve the top of a storm deposit; the term *modification* is here preferred for this process, and it has already been discussed.

Turbulence caused by storms, and gravity are the most common agents of reworking; shells can be entrained by sediment gravity flows at the heads of submarine canyons and along the outer slopes of organic reefs. They are thus laid down at a greater depth with respect to the living place of the organisms. Water motion caused by storms, sometimes with the help of tidal currents, accumulate the skeletal remains in shallower areas (shelf, beach) and in protected water bodies (bays, lagoons).

To understand if, and how much, fossils have been reworked, one can look at their state of conservation: the more fragmented and worn they are, the clearer is the evidence of transport. Attention must be paid, however, to the fact that gravity flows, such as debris flows and turbidity currents,

A   **B**

keep the particles suspended, which reduces the attrition. Delicate remains can thus be preserved in spite of transport distances that are often enormous. Paleontology provides the most reliable criteria for identifying reworked organisms by checking the coherence of the assemblage in terms of age and ecology (incompatible forms are sorted out and become the signature of reworking).   *Photo: G. G. Ori 1970.*

Plate 68
Normal grading

Two sediment cores are here contrasted to show the difference between traction-*plus*-fallout and traction *alternating with* fallout. In the first case (**A**), sand was in suspension with mud, and both were deposited by a turbidity current. There is a small time lag between deposition of coarser and finer materials, owing to hydraulic sorting during fallout, but they are part of the same event. This is testified by the gradual transition from sand to mud, which contrasts dramatically with the sharp, probably erosional base. The sand itself is slightly graded. Vertical grading is called normal when grain size decreases upward. It is the best *indicator* of fallout from a water column. The sharp base and the presence of tractive laminae in the sand provide additional evidence that fallout did not occur in calm water but from a current. In other words, the association of grading and current structures tells us that deposition occurred from a moving, turbulent suspension.

Turbulence is necessary to keep the particles suspended during transport. Coarser and heavier particles need stronger eddies to support them, so when the current looses energy they are deposited first. Finer particles stay longer in suspension and settle after the passage of the current tail. As it is clear that deposition started abruptly in this case, and was attended by waning energy, we are dealing with a catastrophic event.

The turbidite rests on a fine, pelagic mud; compare its structureless character with the lamination in the turbidite mud. Laminae in muddy sediments do not reflect traction, but small changes in settling rate or other local variations (in turbulence, or particle concentration) within the turbulent cloud.

The the second core (**B**) is another example of heterolithic facies, of probable tidal origin, similar to those already described and discussed in plates 60 and 61. Tractive and fallout deposits are spatially separated and form distinct beds; tractive structures, represented by cross-laminae and ripple profiles, are obviously found only in sand beds.

Photos: **A** Institute for Marine Geology, C.N.R. 1992; **B** A. Bosellini 1972.

Plate 69
Normal and inverted grading

Pyroclastic materials form this outcrop, a roadcut in the island of Salina, Tyrrhenian Sea. Two superposed depositional units, separated by an irregular surface, display opposite types of vertical grading, emphasized by coarse clasts (lapilli and lava fragments).

The lower unit is a normally graded bed, faintly laminated in the upper part. The upper unit is clinostratified and reversely graded. Dealing with volcano-derived mate-

rials, one can suspect that normal grading was caused by ballistic fall. If it were so, fine ash would lag behind the coarser particles and be sprinkled on top of them. Instead, they are mixed together, which means that they were suspended and mixed in a turbulent cloud. The effect of turbulence, in fact, is to uniformly distribute the finer materials within the flow, whereas the heavier particles are concentrated near the bottom. When the velocity and

the carrying capacity decrease, the fall of coarse grains entrains part of the fine ash traveling at low levels.

Reverse grading, culminating with some floating blocks on top of the deposit, is a spy to a different mechanism: a laminar, semi-coherent flow with a higher concentration of solid particles. This is a *pyroclastic flow,* i.e., a hot, gravity-driven suspension deriving directly from the volcanic vent or from a convective gas column, or a cold debris flow, wet or dry. Turbulence does not develop, or is choked, when viscosity is high, and heavier particles are supported by the matrix strength (cohesion). Macroscopical characters are not sufficient to reveal whether the flow was pyroclastic or not: textures and structures are the same in both cases. One must look for some specific *indicators,* such as the presence of vegetal debris (if it is charred, the flow was hot) and small aggregates of ash called accretionary lapilli (see plate 166), which are fragile and easily destroyed when the ash is remobilized. For more information on these processes, see introductory section.

ENDNOTE

1. R. G. Walker. 1975. Generalized facies models for resedimented conglomerates of turbidite association. *Bull. Geol. Soc. Amer.* 86: 737–748.

Longitudinal section of a turbidite layer, subdivided into a sand bed and a mud bed. Vertical and lateral changes of thickness, grain size (dots), and sedimentary structures are emphasized. They can be summarized by the terms vertical and lateral grading. **a** through **e:** divisions or intervals of the "Bouma sequence": **a** grain size grading (normal), **b** lower plane-parallel lamination, **c** ripple (small)-scale cross-lamination, **d** upper parallel lamination (plane to wavy), **e** structureless pelite.

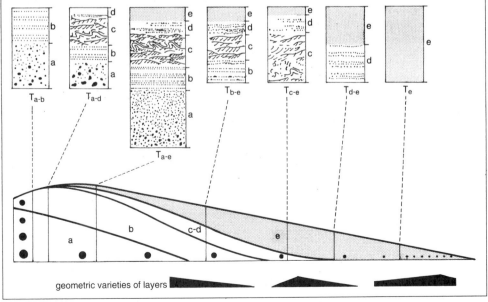

CHAPTER 5
Erosional Structures

Plate 70
Erosional surfaces

Erosional surfaces, more than other structures, pose two problems: 1) how to identify them; 2) how to interpret them.

Identification: in this outcrop of pyroclastic deposits, three discontinuities can be recognized in the vertical succession of beds. Two of them can be seen quite easily because they truncate underlying beds and show minor irregularities. The lower one is covered by discordant (onlapping) beds, the upper by conformable beds. The third surface is more elusive, being almost parallel to bedding surfaces. It marks the top of the darker bedset. Local evidence of truncation can be detected along it by careful observation.

Summarizing, erosional surfaces can be recognized by looking for anomalous contacts, such as angular unconformities, *below and above* them, or for minor morphological features *along* them. A basic rule is the following: bedding planes never cross a discontinuity surface; they terminate against it either from above or from below or are parallel to it.

Interpretation: bed truncation is an obvious indication of erosion. A fundamental point is understanding, if possible, whether the erosion has been synchronous or diachronous. Synchronous means almost instantaneous, or anyway related to a single event, usually a catastrophic one. Diachronous means that erosion occurred in discrete, successive steps over a geologically significant time span. Erosion surfaces due to the cumulative effects of channel migration, as we have seen (plate 15), are diachronous. A diachronous surface must have been exposed to air or water before being covered by sediment. It is also likely that deposition occurred stepwise on it, and was diachronous too.

Although there is some geometrical evidence helping us to decide whether erosion was synchronous or diachronous, the key argument consists in dating it by stratigraphic methods. What can be dated is not the surface itself but the beds lying immediately below and above it. If the time difference is negligible, then the erosion was rapid (short-lived). If not, the surface is diachronous, and erosion lasted for part of the measured time gap (another part is represented by the canceled stratigraphic record). Unfortunately, the normal resolution of geologic clocks, especially in clastic sediments, is too poor for this test to be significant.

Pleistocene pyroclastics, Salina, Archipelago, Tyrrhenian Sea

---

Erosion produces a varied relief with all scales of roughness. Specific features, or elements, of these surfaces are what can be called *erosional structures* (channels, scours, etc.: see plates 73–77). It is not easy, however, to define and pigeonhole morphological characters that often grade into one another.

At the largest scale, erosional surfaces are complex and constitute traces of fossil landscapes. In stratigraphic sections, they have the character of *discontinuities*, or breaks in sedimentation if the succession is entirely sedimentary. Remember, in this respect, that all bedding surfaces are discontinuities, and we do not know in advance the entity of the temporal gap across them. This implies that *every* surface of discontinuity must be regarded with suspicion: it could conceal an important *hiatus*, even if it does not truncate older beds at a high angle or is not accompanied by traces of weathering, paleosols, etc.

Plate 71
Surface of synchronous erosion

A set of sandstone beds is sharply cut by a low-angle discordant surface showing terraced scours in the central part of the picture. A huge sandstone bed overlies the discontinuity: it is not composite, i.e., the product of amalgamation of several beds, but an individual depositional event. A turbidity current of high volume accumulated it. As the volume (or the thickness) of a gravity-driven, density current is one of the factors controlling its energy and velocity, we can assume that this high-volume flow had an erosional capacity.

Having thus recognized an erosional surface that truncates various beds but is mantled only by one, and one deposited by a flow of high energy, we can infer that the same event was responsible for both erosion and deposition, and that the time gap between them was minimum and geologically insignificant. Actually, most surfaces of instantaneous erosion can be accounted for by catastrophic events, both in turbiditic and other systems.

Upper part of Marnoso-arenacea Formation, Santerno valley, northern Apennines.

One further remark about dating rocks and sediments: every method invented by stratigraphers and geochronologists to this purpose has its own limits of precision, or *temporal resolution*. Let us assume that this limit is 10,000 years. If a sequence of beds has employed 3,000 years to accumulate, the depositional and erosional events we recognize in it cannot be dated and are apparently synchronous. They can only be placed in their relative order of succession. This makes the difference between a *cardinal* (truly quantitative) concept, or measurement, and an *ordinal* one (order or sequence of things): a difference that must be clear to geologists and non geologists as well.

An actualistic approach (monitoring erosive processes in the modern world, recording their rates and tempos, and extrapolating data back in time to interpret erosional surfaces) has a limited usefulness in this respect for two main

reasons: 1) the Recent record is too limited in time, and exceptional events have a return time that exceeds it; 2) this period is rather exceptional with respect to the geological past (the climate is colder and more continental than usual, many young mountain chains dominate the landscape, etc.). It cannot be regarded as representative of most geological history.   *Photo: R. Biscaretti 1970.*

Plate 72
Erosional surface at base of a crevasse deposit

This outcrop shows red colored continental deposits (red beds) belonging to the Devonian Old Red Sandstone Formation of NW Europe. Parallel bedded siltstones constitute the greatest part of it. An isolated bed of pebbly sandstone is intercalated; its thickness changes laterally, the top is flat and the base oblique to the general bedding attitude.

All beds represent fluvial sediments deposited outside channels: the siltstones are overbank deposits, settled from suspension in water that invaded a floodplain after spilling over levees and banks. The sandstone is too thin to be the fill of a channel; it is, more probably, a crevasse splay, i.e., bed load sediment that got out of the channel through a breach in a levee (see plate 57). Its emplacement was preceded by scouring of overbank deposits; the basal erosion surface, plus the presence of pebbles in the deposit, suggest that the channel was not too far for the current to have still a high energy.

Notwithstanding the difference in scale with the previous example, the interpretation is basically the same: a catastrophic event (here, a fluvial flood) produced an erosional surface, which was almost immediately mantled by sediment that the same flow transported. Moreover, synchronous erosion occurs at the base of *unconfined flows,* and has nothing to do with channel erosion. It can be qualified as unconfined, or *sheetlike erosion.*

In the case illustrated by this outcrop, in Dunmore East, Ireland, the erosional surface was not smooth: it reveals a small-scale topography, with local scours indicated by lateral thickening and thinning if the bed. The *scour-and-fill* mechanism implies that such surfaces do not remain exposed and cannot be examined in their entirety. It belongs, moreover, to the *smoothing* (or drowning) model of sedimentation, in contrast to the *draping* model, which allows a recognition of the buried topography (see plate 58).   *Photo: G. G. Ori 1992.*

Plate 73
Erosional surfaces: V-shaped channel form

An erosional surface assuming the profile of a channel section is a *channel form*. I avoid the term "channel" without specifications for the following reasons: 1) the geometry of a channel is not always recognizable in stratigraphic sections; for instance, the longitudinal profile can be parallel to bedding with no apparent truncations, and the transversal section is also bed-parallel or complex in case of lateral migration or abrupt shifting (avulsion); 2) channel-looking profiles, on the other hand, can be simple ephemeral scours made by non channeled flows. *Scours do not contain the flow that produces them, whereas channels do*. A single, unconfined flow can simultaneously produce several scours along its base. Furthermore, a channel is a geomorphic and hydraulic unit: except in extreme climatic conditions, it represents a relatively stable element of the landscape and conveys many flows before being abandoned.

In this and in plates 74 and 75, we are going to examine some channel forms displayed by pyroclastic deposits in the Salina islands, Eolian archipelago.

The cliff seen in plate 73 is about 6 m tall, and shows fall deposits of various sizes in distinct parallel beds. The beds are cut by a V-shaped surface, complicated by steps and small terraces. The lower part of the V is filled by deposits similar to the surrounding ones; they are bedded, with a slight inclination. Thin-bedded ash constitutes the upper and widest portion of the fill, and drapes over the shoulders of the V thus "sealing" it. The outcrop is covered by weathered crusts along a flank of the V.

Water running down a steep slope incised this erosional form, which can be called a gully, rill, or *arroyo:* vertical erosion (entrenchment) was favored by the loose state of the materials. Water-carried, reworked pyroclastics were abandoned on the bottom of the V. The rest of the section remained open for a while, until a hot, surge-type flow arrived with a suspended load of fine ash. The ash was deposited in both parallel and cross-bedded units, with a broad concave-up attitude and a meniscus-like contact with the V profile.

This channel form is interpretable as a true channel because: 1) it is not synchronous but was created in several steps; 2) it served for different processes; and 3) was filled by many depositional events. There is no evidence of lateral migration.

Other (not erosional) features visible in the outcrop are the isolated wavelike form (dunoid) near the bottom, with the light band marking a drape of fine ash, two adjacent pits left by the landing of volcanic bombs on another bed of fine ash (near the center of the picture: compare with plate 121), and some vertical structures due to escaping of hot gas (gas pipes) in overlying beds.

Pleistocene pyroclastics, Salina island, Tyrrhenian Sea.

Plate 74
Rounded channel form (scour-and-fill)

Another example of channel form in pyroclastic deposits is emphasized by the strong contrast between the filling materials (light and well bedded) and the substratum (structureless, coarser and darker) in this cliff, about 3 m high. The erosional profile is simpler, more rounded, and asymmetrical in comparison with the previous case (plate 73). The filling is monophasic, i.e., made of only one type of materials with a conformable bedding. Moreover, these materials are not reworked from the substratum; they look quite different from it, and are well sorted. They lap on the flanks of the channel form with the meniscus-like attitude already seen in the upper part of plate 73.

The eroded deposits are thick *lahar* beds, i.e., ash and cinder of fall origin that slid down volcano slopes; sliding or laminar flow produced a chaotic admixture. The laminated fill deposits are, instead, primary; they were emplaced by a turbulent, tractive pyroclastic flow of the ground surge type. Such a flow has a power to erode, and most likely produced both the channel form and its filling in a single event. The structure is thus interpretable as a scour-and-fill. The rounded profile of the scour was carved by eddies of the highly turbulent flow; it is similar to that of smaller structures known as flute marks (see plates 87–91). As for the asymmetry of the profile, it should be the result of centrifugal effects induced by local curvature of the flow path (directed toward the reader).

Pleistocene pyroclastics, Salina Islands, Tyrrhenian Sea.

Plate 75
U-shaped channel form

We are in the same setting as plate 74, with the only variation represented by the shape of the scour. This is a rare occurrence, anyway, because the structure is visible in three dimensions (and has the curious aspect of a big tongue, reminding one of papier-mâché decorations on carnival floats).

The causal agent of erosion is again a pyroclastic surge or a "nuée ardente" (burning cloud) flowing down a steep volcano slope (the slope is still the original one: see 75 **A**). The fill is multilayered, which would suggest a relatively permanent channel; it cannot be excluded, however, that the depositional events succeeded one another in a rapid sequence, possibly during the same volcanic eruption. This seems to be true, at least, for the conformable bedset shown by the longitudinal section (**A**); below it, there is a thin pavement of ash conformably mantling the scour bottom. Actually, we have here a two-phase filling, separated by a discontinuity (downlap); then, there could be a time gap between phase 1 and phase 2. The second and main phase was probably continuous, and shows an example of retrograde filling, or *back-filling,* as indicated by the upslope dip of the beds. Mark these *backset beds,* as their occurrence is very rare.

The U-shaped section can give us further suggestions: this profile is typically produced by laminar flow of a viscous fluid, from lava flows to glacier ice. In sedimentary environments, cohesive debris flows are mainly responsible for it. It is thus possible, in our case, that a high-concentration pyroclastic flow or a cold lahar scoured a channel, which was filled later on by more dilute surge flows.

In conclusion, the channel form illustrated here could represent either a scour-and-fill or a channel. There is no overwhelming evidence in favor of one or the other case.

Plate 76
Nested scour-and-fill with pelitic plug

I have said before that scours are typically made by high-energy, catastrophic flows invading normally quiet environments. It is consequently common to find fossil scours on top of, or embedded by, fine-grained (pelitic) sediments, and filled by coarse materials. An opposite example is documented by plate 77. A sharply defined, troughlike surface truncates a sandstone bed and is covered by a silty pelite.

How can one explain this inversion of what is expected? There are several possible explanations: 1) the current had erosive capacity but did not transport coarse sediment because this was not available; 2) a bed load was present but bypassed this spot and was abandoned down current from it (we also say "in a more *distal* position"); 3) alternatively, coarse materials stopped in a more *proximal* position and occluded the scour. In case

of occlusion, which could occur also in case 2, the water becomes still and mud can settle.

The stratigraphic unit and the depositional setting are the same as in plate 72. The sandstone bed with the scour on top can be interpreted as a crevasse deposit. If you look carefully, you can see another concave surface below the marked one, just above the pencil placed for scale. The actual base of the scour is that one. A third scour surface appears on the left: it truncates the lower one and is truncated by the upper scour. In chronological order, it is the second scour. In conclusion, there is here a stack of concave erosional surfaces. Notice that I have applied the *principle of intersection,* a fundamental tool for establishing temporal sequences of geological events.

Devonian Old Red Sandstone, Dunmore East, Ireland.

Photo: G. G. Ori 1992.

Plate 77
Scour-and-fill in amalgamated turbidite beds

Going down in scale, erosional surfaces look like accessory features or decorations of beds, i.e., as sedimentary structures in the more traditional sense. For convenience, the scale will be qualified as *medium* (meso) when the objects are of the same order as dune bed forms and related cross-bedding, *small* in the range of ripples and related cross-lamination.

This photo was taken in the same outcrop of plates 8, 54, and 86, i.e., in sandy turbidite facies of the upper Marnoso-arenacea Formation in northern Apennines.

The cut is almost parallel to the paleocurrent direction, as indicated by the foreset laminae in the cross-laminated portion of the lower bed. The upper "bed" is actually composed of two beds, amalgamated along a diagonal curved surface that represents the longitudinal section of a scour. This erosional surface is locally marked by mudstone clasts. The scour fill consists of crude laminae dipping down current, a sort of small-scale prograding delta; it is covered by cross-laminae produced by climbing ripple drift, identical to those of the underlying bed. This interval extends to the left above the amalgamation surface, which becomes bed-parallel. Small pebbles are scattered in the sandstone of the fill, which basically shows the same vertical sequence as adjacent tabular beds with a flat base. In practice, the scour only locally modified the geometry of a regular turbidite layer.

Plate 78
Wave-produced scours

Here again is an example of hummocky cross-lamination (see plate 47) in fine-grained littoral sandstones. The concavo-convex laminaset at the center displays a nice hummock of accretionary laminae resting on scalloped, multiple scour. The scoured surface, and part the overlying laminae, are underlined by shell debris. It apparently shifted from right to left.

Plate 78 is a close-up of a vertical cliff, where no scale could be placed.

Pliocene Intra-apenninic Basin, Zena Valley, northern Apennines.

Generally speaking, coarse particles that pave an erosional surface are called *lag deposits*. They, in fact, lag behind smaller particles that the flows are able to entrain with them, and represent trapped, residual materials. When a stream erodes its bedrock, for example, pebbles can be trapped by minor topographic irregularities. Or, pebbles transported by a high-energy flow, then deposited in a sand bed, are freed (exhumed) by a gentler current that removes the sand.

A lag pavement can be continuous or spotty, and is commonly the result of more than one event. In other terms, it records, in many cases, the passage of several currents, and hence a zone of *bypass* and a time of nondeposition. This implies a time gap of possible geological significance along erosional surfaces paved by lag materials. Scour-and-fill structures have no or little lag deposits, because deposition occurs soon after erosion. Lag deposits are thus features suggestive of *diachronous* erosional surfaces, including paleochannels.

Plate 79
Indicators of intraformational erosion: mudstone clasts

Sometimes the evidence of erosion is indirect only. When a current erodes a muddy bottom, the softer superficial mud is disgregated and dispersed in fine particles within the flow. The underlying, stiffer mud is more resistant and, when it yields, it does in the form of chunks and flakes, which can be incorporated by the current. These fragments, ripped up from the sedimentary environment itself, represent *intraformational,* or *intrabasinal* clasts (other generic terms are *rip up clasts* and *clay (chips).* They can be strewn on the erosional surface as lag deposits, or be carried some distance down current and deposited together with sand or gravel. Intraformational clasts will then be found *within* beds of coarse sediments.

Mudstone clasts are useful *indicators* of intrabasinal, penecontemporaneous erosion: in other words, they point out that erosion occurred at the expense of previous sedi-

ments, possibly not far from the place redeposition. The examples shown here are rounded and platy, and can be described as mudstone pebbles. Rounding can be attributed to bouncing and rubbing, or to abrasion by whirling sand. A flat shape is common, and can be either primary, if the clasts derive from laminated mud, or acquired by compaction under the load of overlying sediments.

Plate 79 **A** is a section view: the mudstone chips are aligned just above an erosion (amalgamation) surface between two turbidite sandstone beds of the Messinian Colombacci Formation in central Apennines. The contact is not well defined, but the two beds can be distinguished because the lower one is laminated (cross and convoluted laminae: see plate 116), and the other is structureless.

Plate **B** is a plan view of a parting plane, i.e., an intrastratal surface. Elongated clasts show variable orientations, with two slightly prominent modes, respectively parallel and orthogonal to the hammer direction. Some pebbles probably rolled along the bottom after being carried in suspension, others did not.

Marnoso-arenacea Formation, northern Apennines.

Photo: M. A. Bassetti 1992.

Plate 80
Indicators of intraformational erosion: armored mudstone clasts

Mud clasts originally have angular, jagged outlines, and keep them if they are carried in suspension; rounding ensues from collisions, i.e., from solid friction between moving objects, when the clasts stay on the bottom and are hit by other particles or are entrained with the bed load. Rolling is the main moving mechanism of large clasts within the bed load. It implies a positive feedback: edges and corners of rolling particles are blunted, which increases "rollability."

If mud clasts reach a sand or gravel bed, sand grains and pebbles tend to stick to the plastic mud thus forming a protective armor. The natural tendency of bed load transport is, in fact, of disintegrating the mud clasts into smaller and smaller pieces (which are actually found in many sand beds) or into constituent silt and clay particles (which will be included in the sediment as matrix). In other terms, mud clasts tend to be destroyed and to disappear after some distance of travel.

The mud clasts shown in plate 80 were armored on the bed of a torrential stream of the Apennines.   Photo: Insolera 1970.

Plate 81
Erosional structures: sole marks

I introduce here a group of structures that display a great variety of forms, often elegant and highly ornamental, at a small to intermediate scale. They develop on depositional surfaces and are thus preserved on bedding planes, especially at the base of sandstone beds: that is why these structures are called sole marks, or sole markings. More often, what is preserved is their mold; consequently, they are labeled as *casts* (see Introduction, figure 5).

Erosion plays a major role in producing sole marks; it can act in more than one way, which results in the morphologic variability we observe. Part of this variability has to do with the physical state and lithology of the substrate. Many structures are carved in mud by catastrophic processes invading a tranquil bottom. They are immediately molded by sand, subsequently cemented by diagenesis into a sandstone; this is the case for the example illustrated in plate 81. Other marks are left on a sand

bed (see following plates); chances of observing them in Ancient sediments are slight because they are more commonly canceled or covered by other sand (the contact is welded and beds do not part along it). Only a mud cover allows them to be exhumed and observed.

The slab seen in plate 81 comes from the base of a sandstone bed belonging to Pleistocene shallow-water "yellow sands" in the Apennine foothills near Bologna. What you see are casts of the original bottom topography (as if you were looking at the depositional interface from below). A scour was made around and behind an oyster valve, fixed to the bottom, by a current flowing from top to bottom of the image. This is called an *obstacle scour* or a *crescent mark,* depending on whether emphasis is on its cause or its shape. A modern counterpart is shown in plate 82.   *Photo: P. Ferrieri 1992.*

Plate 82
Obstacle scours (crescent marks)

On a rough bottom, the boundary flow lines of a current are subject to various disturbances, as we have seen in describing bed forms and pebble clusters. Flow separation and reattachment can occur in many ways. In front of a protruding object, for example, the compression of the flow can cause either a jump of flow lines, with the formation of a shadow zone, or a focusing of energy. Deposition follows in the first case, erosion in the second. When erosional effects dominate, the eddies of the turbulent current impinge both in front of the obstacle and on its sides, forming a lunate, or crescent-shaped scour.

In the case illustrated here, the rear of the largest pebble was protected from turbulence, which favored deposition; the shadow zone is marked by a pebble cluster (see plate 66). On the back of smaller obstacles, only a thin ridge of sand accumulated. The photo was taken on a river bed, after a flood.

On a much larger scale (kilometers and tens of kilometers), similar scours (*moats*) are made by bottom currents around relieves of the ocean floor.

Note, at the bottom, the marks left by the claws of a bird, and compare them with the fossil analog seen in plate 148.

Plate 83
Comet-shaped obstacle marks

Swash and backwash, especially the last, produce these long and shallow scours behind flat pebbles lying on the sand of beach face. The water cover is very thin and does not form moatlike scours around these minimally protruding objects.

The prevalence of platy pebbles is an effect of shape sorting by waves (compare with plate 64). More globular pebbles (spheroidal and "triaxial" shapes) were certainly present in the original gravel carried to the sea by a river. Being less adherent to the bed, they were more easily removed from the foreshore and carried under water, also because gravity helped the backwash.

Photo: E. Rabbi 1992.

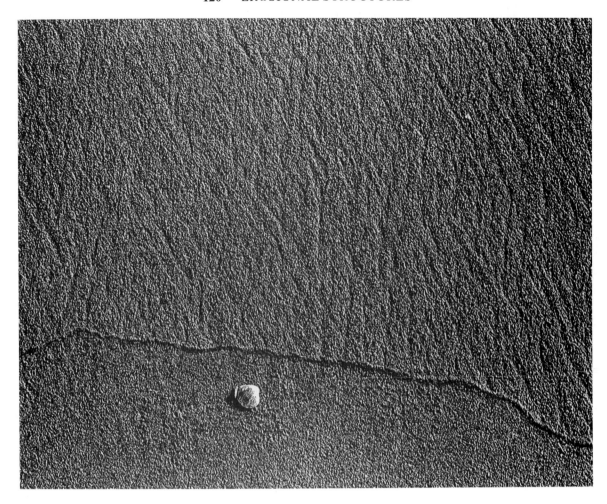

Plate 84
Delicate scour marks on beach sand

These structures have a very shallow relief and a low preservation potential. Again they are found on the swash zone of a sandy beach, a hydrodynamically smooth surface that is permanently in upper flow regime. Part of the visible surface is smooth indeed (the shell has been placed on it for scale), and represents an area where the front of an incoming wave died. Another part is a little rougher, with shallow striations crossing at acute angles. The striations were made by previous waves, in particular by the interference of swash and backwash: when wave fronts are oblique to the shoreline, they produce oblique marks in climbing the beach face. The return flow is controlled by gravity and follows the maximum dip, which is normal to the shoreline. Under the repeated push of oblique waves, solid particles follow a sawtooth pathway and are drifted along shore.

The boundary between the smooth and the rough areas is marked by a narrow ribbon of sand, a *swash mark*. The wave front dies there and abandons the sand it carried. Swash marks form a pattern of arcuate, intersecting lines whose concavity faces seaward in fair weather conditions. Coarser and heterogeneous debris, including flotsam and jetsam, is accumulated by storm waves in arc-shaped berms.

Delicate marks are imprinted on the sand by the wind, too. The sand is dry in this case, and the wind uses the ends of fixed plants (grass blades or twigs) as tools, producing arcuate and circular scratches (*swing marks*, see color photo 22). When wind gusts slow down or stop, the flexible plants go back to their rest positions, and in so doing emphasize the grooves. The process can be repeated many times along the same tracks, as in record-playing by a pickup.

Photo: G. Piacentini 1970.

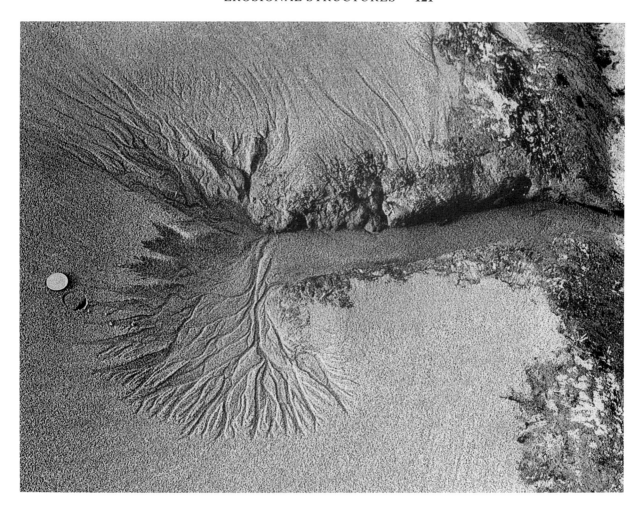

Plate 85
Rill marks: dendritic pattern

Rill marks are narrow grooves with a dendritic or anasto-mosing pattern, often looking like scaled models of river channels. In this case, for example, they simulate two drainage basins joining in a major channel, or, if you like, the head of a submarine canyon. The rill pattern was made by the backwash of a wave that passed over a beach berm (the sea is to the right).

Other varieties of rill marks are illustrated in color photos 19 and 25. They are produced, in general, by thin films of water flowing at moderate to low speed. It seems surprising that such weak currents have an erosional capacity, but we must remember the concept of flow re-gime: a reduced flow thickness compensates for velocity, and can give the current a grip on the bed. Moreover, the water is clean and consequently aggressive: it can pick up grains because its transport capacity, though small, is unsatisfied.

Rill marks and other *minor* erosional structures, parts of which we have already met (see plate 29, for instance), could be included in the group of *modification structures*.

Erosion in subaerial or very shallow water settings is a typical form of modification, and affects both tractive and exceptional deposits and structures.

Plate 86
V-shaped scours

Plate 86 demonstrates the difficulty of recognizing structures on amalgamation surfaces (sand on sand). A turbiditic sandstone bed shows the already seen vertical transition from plane-parallel to cross laminae, in conditions of apparently continuous sedimentation. The lower laminated interval is, however, interrupted by V-shaped scours; several of them are nested (a sort of cone-in-cone arrangement) beside and above the hammer head, and can be laterally traced in at least two amalgamation surfaces.

Other scours are barely visible on the right-hand side. Note that they are absent from the cross-laminated interval.

It is not clear whether the scours were made by different currents or by velocity pulsations within the same flow.

Marnoso-arenacea Formation near Sarsina, northern Apennines.

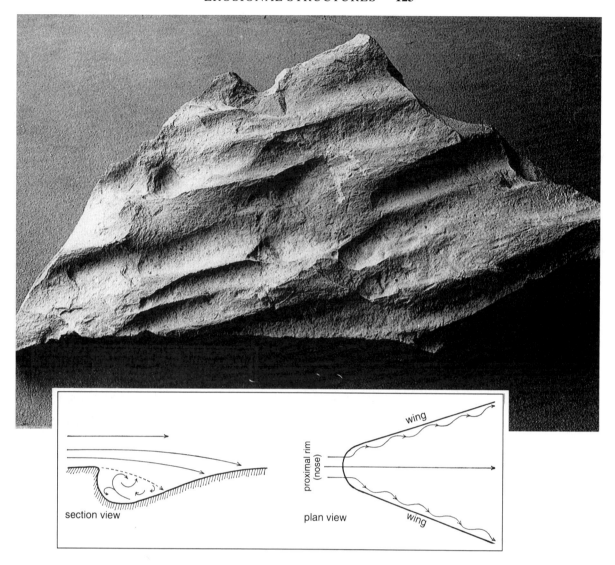

Plate 87
Erosional marks on pelite: flutes

These flute-shaped furrows are parallel to the flow, with an upcurrent (proximal) nose and a distal open, transitional end. They are usually found as casts on the sole of sandstone beds, as the underlying mudstone is more easily weathered. The specimen of fresh mudstone photographed here is rather special, and was collected in a quarry after mine blasting.

Marnoso-arenacea Formation, northern Apennines.

Flute marks can occur isolated, but are more commonly clustered or cover the whole bed surface, as the following pictures show. The scouring agent is a turbulent current (the marks are frequent at the base of graded beds), with local eddies focusing their energy on or behind small irregularities of the bottom (see inset). In other words, flute marks should be a variety of obstacle scours, even though no fixed objects are associated with them. On a deep-water bottom, one cannot expect pebbles or stones, i.e., the type of roughness that can rather characterize river beds. Local asperities on a mud bed can be provided by the fill of animal burrows (see plates 151, 156), and local zones of stiffened mud.

Photo: G. Piacentini 1970.

Plate 88
Fluid-made marks: flute casts

The base of a verticalized turbidite bed is shown here (see the commentary for plate 9). It is completely covered by flute casts, from which the paleocurrent can be easily defined (try yourselves, on the base of what has been explained in plate 87) and measured. A hammer is placed for scale in the lower left side.

Marnoso-arenacea Formation, northern Apennines.

Flute casts are the most "popular" and widespread markings in turbiditic sandstones, where they represent the most reliable *indicators* of paleocurrent. It would be wrong, however, to consider them as *indicators* of turbiditic or deep-water sedimentation. A turbulent suspension flowing on a muddy bottom is all you need to form flute marks. Such a current can occur in shallow water by means of tidal or storm input, and also in the alluvial environment; flutes have been found in lagoonal sediments, for example, and even in bedrock clays of some stream beds and banks (see plate 148).

A

B

Plate 89
Flute cast morphology

Flute casts are aligned in rows parallel to the flow in these examples deriving from two different turbiditic formations of the Appenines. Blunt-nosed cylindrical forms are associated with more pointed and conical forms. Some scours are deeper than others and show "twisted" parts probably reflecting the action of helicoidal vortexes.

The two outcrops are typical of tectonically deformed zones in fold-thrust belts, with the sandstone beds standing out amid more erodible mudstones.

A: *Marnoso-Arenacea Formation, Miocene, Umbria region;* **B:** *Monte Sporno Formation, Paleocene Bagamza Valley, northern Apennines.*

Plate 90
Flute and crescent casts

A variety of flute morphology is shown by this slab of sandstone. The preservation of morphological details by the molding sediment can also be appreciated.

The parallelism of flute casts is impressive. Two oblique traces are visible, but were caused by different mechanisms. One is to the left of the pencil, and is a tool mark (a scratch made by an impacting or dragged object), the other (near bottom center) was left by a burrowing organism. The burrow is cut into segments by flutes, which implies that it was preexistent to scouring.

Scattered among the flute casts, are small, elongated crescent casts (see plates 81–83), suggesting that tiny obstacles protruded from the bottom. Some of them could have been brought by the eroding current itself. Nowhere was the bottom perfectly smooth; numerous small marks studded it along with the flutes.

Marnoso-arenacea Formation, northern Apennines.

Plate 91
Clustered flute casts

A vertical bedding plane is a vantage point for observing sole marks and their areal distribution on Ancient bottoms. In contrast to plate 90, flute casts have here a larger size and a flatter shape (as can be measured, in relative terms, by the ratio between width and depth), with broad, blunted noses.

Moreover, flutes tend to cluster in bands or stripes parallel to the paleocurrent; the depth of scouring is greater in these bands than in intervening areas. This suggests a hierarchy of erosional forms, with the flutes forming minor features of larger scours; you can find a similar relation in tractive bed forms, with ripples printed on the back of dunes and dunes superposed to larger forms.

Marnoso-arenacea Formation, northern Apennines, Monte della Faggiola.

Plate 92
Gutter casts

Gutters, as the name implies, are scours with a more cylindrical and elongated shape as compared with flutes. Their surface can be made irregular, in detail, by minor grooves and asperities, as the example shows (unfortunately, in section view only).

Occurrences and causative mechanisms are similar to those of flute casts. At the center of the picture, we have here a fluvial crevasse deposit interbedded in continental fluvio-lacustrine clays. It is a single depositional event, starting with a conglomerate and ending with a laminated sandstone (needless to say, a catastrophic event).

Part of the pebbles and sand grains are made of white, alabastrino gypsum; this underscores that evaporitic minerals and crystals can be removed from the marginal areas where they precipitate, when the water levels drops because of protracted evaporation.

Messinian deposits of Tuscany, not far from Volterra, the type locality of alabastrino gypsum in Italy. Gypsum alabaster is quarried and used as ornamental stone.

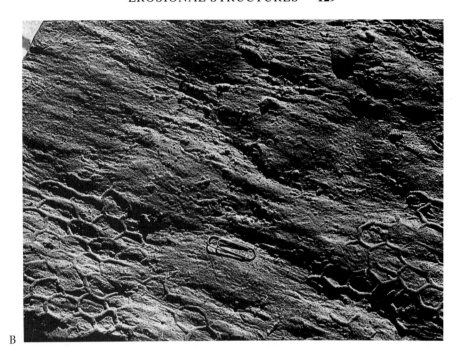

B

Plate 93
Delicate erosion marks

As in plate 90, we see here that tiny markings can fossilize very well. They can be either associated with larger forms or represent the only structures of a bedding plane. The latter case occurs more frequently in thin beds of fine sandstone and siltstone, and reflects the fact that the current had lost energy and erosional capacity. Smaller eddies thus produced smaller markings. This is not always the case, however. Delicate striations and scratches can be produced also by vigorous flows meeting small asperities of the bottom or dragging sand grains over it (see also tool marks, plates 98 to 103).

Delicate marks, seen in proper light conditions, are no less useful than flute casts in detecting the paleocurrent direction. The picture shows how clear their parallelism can be, plus the fact that they overprint larger forms of subdued and unclear morphology. Biogenic cellular structures (*Palaeodictyon,* see plate 159) were here stretched and torn by the same current that made the marks, a clear evidence that the organic trace was present on the mud surface before the arrival of the current.

Marnoso-arenacea Formation, northern Apennines.

Plate 94
Rib-and-furrow and pinch-and-swell structures

This oblique view shows a hummocky bed surface. At first sight, one is inclined to regard this morphology as the interfacial expression of real hummocky cross-bedding (see plates 47, 78) or as a rippled bed top. Actually, it is the base of an overturned sandstone bed, as indicated by the small sole markings, and the hummocks are molds of depressed portions (see inset). They reflect the morphology of current ripples, in particular of ripple troughs that partly sank in the underlying mud. The upper bed surface, not visible, is undulated, too, as indicated by the sketched section; the opposite undulations give the bed a geometry that has been named *pinch-and-swell*.

The arcuate lines visible in the lower right and upper left are intersections of foreset laminae with the sandstone sole. The bisectrix of the arcs, toward the concave side, gives the paleocurrent direction (compare with plate 35). As you can check, it coincides with the direction of sole marks. The sets of arcuate lines are known as *rib-and-furrow* structure.

Turbidites of the Marnoso-arenacea Formation, northern Apennines.   Photo: G. Piacentini 1970.

Plate 95
Dendritic ridge casts

These markings allow us to recognize the paleocurrent direction, which is the direction of convergence of the dendritic lines. The lines look like fissures but are actually the molds of narrow ridges of mud alternating with wider and flatter furrows. As the inset shows, this minor topography was created by a current flowing on a muddy bottom: at the base of the current, cylindrical vortexes would form, with water particles rotating in them to give a helicoidal flow. The sense of rotation would alternately change in adjacent vortexes (called also *spiral tubes*), which are organized in couples as shown by experiments. Each couple pushes the sediment apart where descending flow lines diverge, and suck it up where ascending lines diverge. Furrows and ridges would thus be made at the same time, as parts of a single process that combines erosion with deformation. Were the mechanism purely erosional, the crests would be residual features, i.e., non-eroded parts of a more elevated surface.

The structure has been reproduced in flume experiments, verifying that tubular vortexes tend to converge down current, and that where convergence actually occurs, a coupled vortex is lifted and looses contact with the bed.

In section view, mud ridges are pointed and can deviate from the vertical; they virtually cannot be distinguished from *flame structures,* produced by load deformation and upward squeezing of mud (see plates 123, 124, and color photos 7 and 8).

Macigno Formation, Lower Miocene, northern Apennines.

A

B

Plate 96
Current markings on lamination surfaces

Shown here are fissile sandstones split into individual laminae or laminasets, whose surfaces can show delicate markings parallel to the paleocurrent. Splitting occurs preferentially along surfaces of weakness, where fine or lamellar particles (mica, vegetal debris) are often concentrated. These *intrastratal* markings were probably produced by current drag in a way similar to that responsible for the structure seen in plate 95. The tubular vortexes were smaller in this case, and confined to a thin layer of water close to the depositional interface. The marks are also more parallel than dendritic types, although some convergence ("fleur-de-lys pattern") can be seen in 96 **B.** The furrows are interrupted by arcuate septa that point to the current direction and suggest a certain resemblance with flute casts.

These two examples do not adequately document the considerable morphological variety of intrastratal marks. A comprehensive term proposed for the structure is *longitudinal furrows and ridges.*

The examples shown here derive from turbiditic sediments, in particular from the **b** division (plane-parallel laminae) of the Bouma sequence (see plate 69, inset). As we know, in turbidites, laminae record tractive effects of a waning current; contrary to basal marks, which reflect the current direction *before* it starts deposition, intrastratal marks are produced *during* deposition. Their value and utility, as *indicators,* consist in allowing us to check whether the current kept its initial direction or made some turn or deviation while depositing its sand load.

Marnoso-arenacea Formation, northern Apennines.

Photo: G. Piacentini 1970.

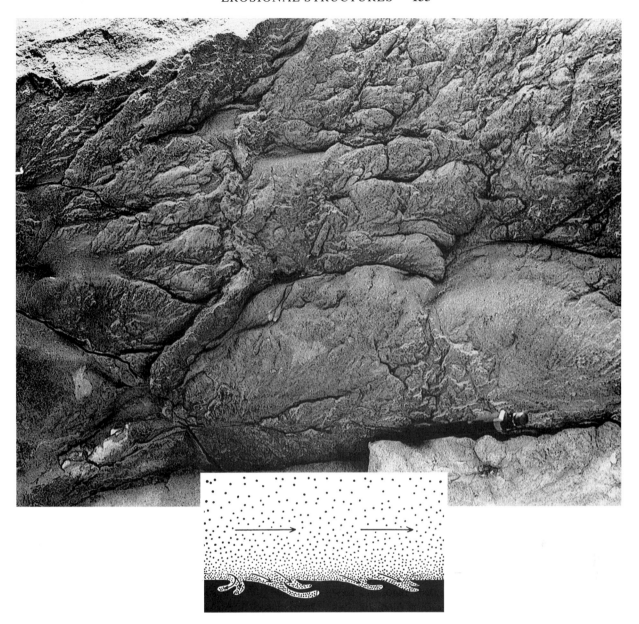

Plate 97
Frondescent casts

The shape of these curious, lobate markings strongly reminds one of cabbage leaves. They derive from sand filling small "cul-de-sacs," blind fingerlike furrows with a diverging pattern (see inset). Note that this pattern is opposite of that of dendritic ridges (plate 95); a distinction among the two structures is thus essential. Frondescent casts can be recognized by details of their ornamentations. Here again we see, in the form of fissures, the pelitic crests, or "flames," that separate the sand lobes.

The structure should be another manifestation of combined erosional and deformational effects of the basal portion of a density current. A high concentration of sediment or a particular form of turbulence may have helped in underscouring the bottom. It is difficult to say whether the basal layer of the current was in laminar or turbulent conditions; in the former case, the fluid penetrating the mud bed was more likely dense quicksand rather than water, whereas some sorts of vortexes were the undermining tools in the latter. Whatever the case, the fluid had certainly a high density, whereby one could put these marks in a category of *fluid impact* or *fluid drag* features (see also plate 99).

Macigno Formation, northern Apennines.

Plate 98
Tool marks: groove casts

The term *"tool mark"* refers to a trace of erosion made by an object carried by the current rather than by the fluid itself. In this category, various forms are distinguished by specific names. In this plate, for example, *groove casts* are illustrated by a magnificent exposure in overturned turbidite beds. The sole was exhumed when a forest road was open, but rapidly deteriorated: the picture already shows the progressive detachment of sandstone slabs from the almost vertical wall, reminding of the unlucky fate of frescos abandoned to pollution.

Grooves are linear, parallel marks, of uniform relief, locally intersecting at acute angles (lower than 40°). Their surface is either smooth and rounded or striated. Sizes are extremely variable. They are produced by solid objects impacting on a mud bottom and dragged by a current for several decimeters or meters. These objects consist of mudstone fragments, shells, or waterlogged plant fragments (twigs, branches, etc.); they were present in the source area of the flows or became incorporated by intrabasinal erosion. Tools are rarely found on the bed surfaces, even where the ends of the marks are visible. This is not remarkable, however: materials like clay and vegetal matter, for example, are deteriorable and tend to disintegrate owing to friction with the bottom; those that survive can be deposited within the bed, as in the present case where clay chips occur a few centimeters above the base (see plate 99).

The fact that the objects slid on the bottom, with little or no rotation, suggests that they were too heavy to be lifted by eddies, or that turbulence was suppressed near the base of flow by a high concentration of suspended particles. Eventually, however, they were raised in the fluid, as they are not found at groove ends (see inset).

Marnoso-arenacea Formation, northern Apennines.

Photo: G. Piacentini 1970.

Plate 99
Groove casts and fluid drag casts

More details of groove morphology can be detected in these partial views of the previous outcrop. Beside the compresence of smooth and striated groove casts, produced by rounded and jagged objects, respectively, you can see the occurrence of lobate marks, partly similar to the frondescent forms shown in plate 97 (*fluid* impact and/or fluid drag structures, also called by someone *flowage casts*). Their down flow divergence enables the deter-

mination of the paleocurrent direction, which grooves alone do not indicate.

Along the intrastratal surface exhumed by peeling off the basal "crust," mudstone pebbles occur, isolated or grouped (see upper right-hand corner). These are the "tools" that made most of the marks.

Marnoso-arenacea Formation, northern Apennines.

Plate 100
Pebble-made groove casts

These marks were photographed on a detached block of pebbly sandstone, and have a very pronounced relief. Rounded groove casts are recognizable, part of which has a sinuous trend. They end in, or are overprinted by bulbous, lobate mounds that conceal pebbles, visible within the bed in sections (very few of them are preserved on the basal surface). The weight of the pebbles made the sand sink into the mud; this load effect is responsible for the roughness of the surface.

The sandstones are not classical turbidites (as were most examples presented earlier) but are rather proximal gravity flow deposits of a fan delta. The closeness of a fluvial source explains the abundance of pebbles.

Colombacci Formation, Pietrarubbia, northern Apennines.

Plate 101
Marks produced by dragging objects: chevron cast

Chevron, or skim casts are made by objects that graze the bottom but do not sink into it (in anthropomorphic terms, they tickle the mud). The response is an oscillatory deformation similar to the waves produced by a hull on a water surface. The resulting V-shaped wrinkles are ideal *indicators* of the paleocurrent direction (they point down current).

The example shown here, from the Paleozoic Aberyst-wyth Grits of Wales (U.K.), represents the local transformation of a groove. The grooving object, coming from the right, was raised a little, produced the chevron mark, then sank again reforming and extending the groove.

The rest of the bed surface is fairly smooth, exception made for scattered marks of squat shape, deriving from the momentary impact of bouncing and skipping objects.

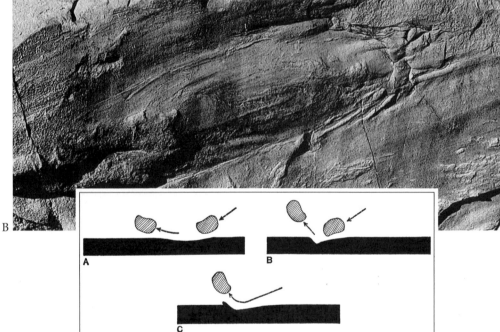

Plate 102

Marks produced by impacting objects: bounce casts, prod casts, brush casts

The small groove casts and *striation casts* in plate 102 **A** are associated with short marks due to impacting tools. A certain deviation from parallelism can be noted in both groups; it can be attributed to statistical fluctuations around the mean direction of a single current, to winding paths of flow lines, irregular whirls, etc. Angles greater than 40° would suggest, instead, the passage of more than one current, and sedimentologists would face the problem of deciding which system was the "good one" for indicating the paleogeographic trend.

Three main types of *impact casts* can be recognized: 1) *bounce casts* are symmetrical in longitudinal section (inset A) and, consequently, do not allow determination of the current direction; 2) *prod casts* (inset B) are asymmetrical, with a more pronounced relief on the down current

end (contrary to flute casts); 3) *brush casts* (inset C, and plate 102 **B**) are similar to bounce or prod casts but the frontal end is marked by one or more wrinkles due to mud dragging.

Try do identify the different types in 102 **A** and, if possible, the current direction. Pay attention to the similarity between prod casts and flute casts, which give opposite indications. In this respect, remember that prod casts commonly occur as isolated marks, while flutes rarely do. Moreover, tool marks and flute marks are frequently (but not always—see plate 103) mutually exclusive or almost so. On many bed surfaces, you will find either one type of structure or the other.

Slabs from sandstone beds of Marnoso-arenacea Formation, northern Apennines.   *Photo:* **A** *P. Ferrieri 1992.*

Plate 103
Association of tool and fluid-made marks

Another case of bed turned upside down, this time in a Cretaceous flysch (Monghidoro Formation, northern Apennines), displays finely chiseled basal markings. Subdued flute casts are associated with a variety of shallow scratches (groove, striation and impact casts). In addition, they are overprinted by short, sinuous, keeled marks made by organisms (trace fossils: see plates 156, 157). The paleocurrent direction is determinable. Can you do it?

Deformation of sediments can begin *during* deposition and be caused either by the depositional process itself or by other mechanisms triggered by it or by gravity. Deformational events also occur *after* deposition on or below the sedimentary interface. A first distinction can thus be made between *syn-* and *post-depositional* deformations, with the proviso made in the introduction, i.e., ignoring tectonic processes. In certain cases, however, it is not easy to discriminate between sedimentary and tectonic causes of sediment deformation, especially if tectonic stresses act on a still soft sediment. Some examples will be shown and discussed to make this point clear.

The following survey of deformative structures does not pretend to be systematic and exhaustive. It is rather aimed at exemplifying common and significant cases. I will start with smaller, gentler, and more localized structures and end with forms of larger scale and greater intensity. At one end of the spectrum, only parts of beds or bed surfaces are involved, at the other end can be seen deformation throughout packets of beds, sedimentary bodies and stratigraphic successions up to mountain size.

Plate 104
Desiccation cracks (mud cracks)

The fine-grained particles of mud retain a great amount of water due to the high porosity of their open fabric. When water evaporates and the mud dries up, a considerable reduction of volume occurs. Shrinking of the sediment does not only result in particles getting closer, but in the formation of widening cracks that separate discrete masses. The cracks cross each other and delimit polygons of mud; the reason for this pattern is that the mud reacts as an isotropic substance (equal response in all directions) to stresses that are mostly horizontal. Polygon sizes and fissure widths are proportional, and increase with the thickness of the drying mud. The cracks are wedge-shaped and close downwards.

In the picture, two systems of cracks of different size are recognizable: the larger cracks truncate the smaller ones. Small cracks affect a thin mud layer; they formed first, and, with progressive desiccation, some of them were enlarged while the others became inactive. In other terms, during desiccation, contractional forces extend their action in depth but restrict their area of application by reducing the number of fissures along which they act.

See footprints for scale.

Photo: V. Rossi 1970.

Plate 105
Desiccation structures: mud cracks, polygons, and curls

It is possible to see, in this close-up view, the relations between the sizes of desiccation cracks and polygons. The edges of the thinner superficial film have been curled; the process goes on until dry mud rolls and curls are completely detached from underlying, wetter mud. Curling is just starting at the edges of larger polygons.

In conclusion, it can be seen that the ultimate result of this deformative mechanism is similar to that of intraformational erosion: it produces mud clasts, which can be picked up by transporting agents and redeposited somewhere else.

Sediments can be fissured by processes other than desiccation. Cracks can form, for example, by dilation rather than contraction. This is the case of *ice wedge cracks:* in periglacial and circumpolar areas, the soil is frozen most of the year. Thawing can occur in superficial layers during the warmer season. It is well known that water increases its volume when freezing; as a result of this expansion, water pipes break in cities; in the field, ice layers form in surface water of ponds and lakes (see color photo 28) or in saturated topsoil, where ice wedges can penetrate to a certain depth. Upon seasonal melting, the ice leaves open fissures, and soil debris and stones can fill them. The cracks can thus fossilize and testify to cold conditions of the past; they are utilized as *paleoclimatic indicators*. Other kinds of deformation occur in glaciated areas; I skip these peculiar structures, just mentioning the term (*cryoturbation*) that encompasses all of them.

Cracks can form in subaqueous mud, too; the mechanism is one of shrinkage, as in the case of desiccation, but the process is clearly different. It is called syneresis, and is well described by the A.G.I. Glossary: "the spontaneous separation or throwing-off of a liquid from or by a *gel* or flocculated colloidal suspension during aging, resulting in shrinkage."[1] Actually, the "spontaneous" separation of water from mud is induced by mechanical disturbances or chemical changes, such as variations of salinity (contents of dissolved salts), etc. Fissures due to this process are called *syneresis cracks*.

Plate 106
Fossil desiccation cracks (molds or fills)

If the mud polygons do not curl up completely, the cracks can fossilize; this happens by means of the sediment that buries them, for example sand. After diagenesis, the filling material can become more resistant to weathering and the cracks are preserved as casts (see inset). Fossil mud cracks are very useful as *indicators* of subaerial exposure; they can be found in both continental sediments and marine or littoral deposits that emerged subsequently. The emergence of a marine environment means a regression.

On the other hand, cracks made in continental environment and filled by marine sediments will indicate a submergence, i.e., a transgressive event.

Fossil desiccation cracks and their molds can also be used as *indicators* of stratigraphic polarity (way-up criterion).

Upper Triassic (Raiblian) red beds, southern Alps.

Photo: P. Ferrieri 1992.

A

B

Plate 107
Tepee structure

The name of this structure derives from the rooflike attitude of deformed and broken beds. It is found mainly in carbonate and evaporitic sediments, in which early diagenesis or direct precipitation results in hardened beds and crusts. A main cause of deformation is the dilatancy of materials, in this case; dilation can be caused, in its turn, by thermal or chemical processes inducing mechanical effects, such as heat exchanges, hydration-dehydration, and force of crystallization of salts among others.

In plan view (**A**), tepees are polygonal structures with raised edges. The pattern is thus similar to that of desiccation fissures, but the relief is inverted. The dilation of superficial beds, or the shrinkage of underlying sedi-

ments, generates compressional forces in the horizontal plane. This happens because the materials are laterally confined and cannot expand freely. Beds "rear up" under compression, and tend to bend. If they are not elastic enough, bending cannot occur: brittle carbonate beds, for example, are ruptured, while plastic, hydrated materials are folded (see plate 104 **B**). In other words, they respond to stress with a permanent, anelastic deformation. In "roof tops," facing edges of broken beds thrust over each other. The formation of these "anticlines" and overthrusts allows dilation to occur, as is confirmed by the voids that are created below the interface.

Tepee structures are found in hot and dry areas, in

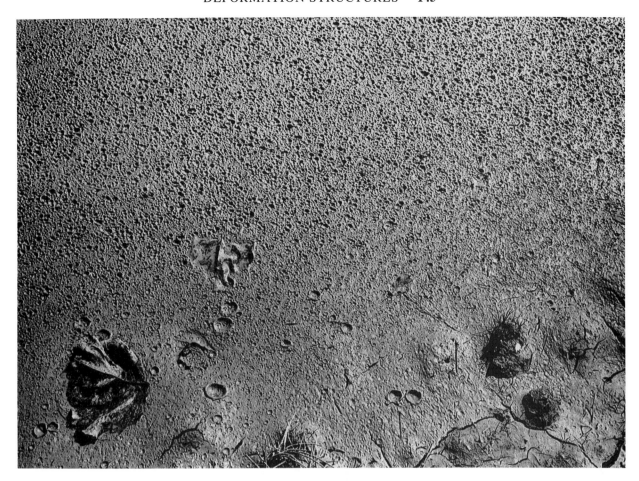

environments like salt pans, *sabkhas, chotts, playa* (ephemeral) lakes and supratidal zones of coastal areas. In the fossil state, they thus represent *indicators* of similar environments and, more generally, of subaerial conditions. The example shown in **B** derives from a Mesozoic (Triassic) unit of southern Alps, the Calcare Rosso.

As early diagenetic processes are involved in tepee formation, the structure might be classified as secondary as well (see last section, plates 162 to 180). Moreover, later diagenetic effects are visible in fossil occurrences; look, for example, at the dark coatings in **B**, representing pore-filling cement crystals. This point stresses that classifications are useful in science as guideline but pigeon-holing of natural objects is difficult and largely arbitrary.   *Photos:* **A** *G. G. Ori 1992;* **B** *F. Jadoul 1992.*

Plate 108
Pitted mud surfaces

Small pits can be produced on wet mud by two distinct processes: the impact of rainfall drops (or hailstone) or the escape of bubbling gas from the sediment. The alternative terms *raindrop marks (casts),* or raindrop impact pits, and *gas escape marks (casts)* can therefore be used to describe the structure. To make a distinction, the morphology of the micro craters does not help much as it is similar in both cases. Their sizes, on the other hand, should be more uniform when rain is the cause. By sectioning the sediment, one can look for evidence of internal deformation due to gas or liquid escape. The example shown here derives from the same area where mud volcanoes exist (see plate 109), and represents marks left by exploding bubbles.

Whatever the case, this relatively rare structure is to be placed together with the other *indicators* of emergence, or subaerial environment. It can be preserved quite easily in dry mud.   *Photo: D. Insolera 1970.*

Plate 109
Injection structures: mud volcanoes

Natural gas can escape from leaking reservoirs in the subsurface or decomposing organic matter at the bottom of lakes and swamps. It can get out dry or mixed with saltwater and fine sediment. In the second case, cones of mud are built around vents. Plate 109 documents a *salse,* i.e., the initial stage of cone formation, which consists in successive additions of lobes and tongues of mud made by small-scale mud flows. The wrinkles you see are "frozen" when the mud stops and dries out; another morphological characteristic of mud flow deposits is the steepness of frontal and lateral edges, due to the cohesion of the mud.

Salses of Nirano near Modena, Italy.   Photo: D. Insolera 1970.

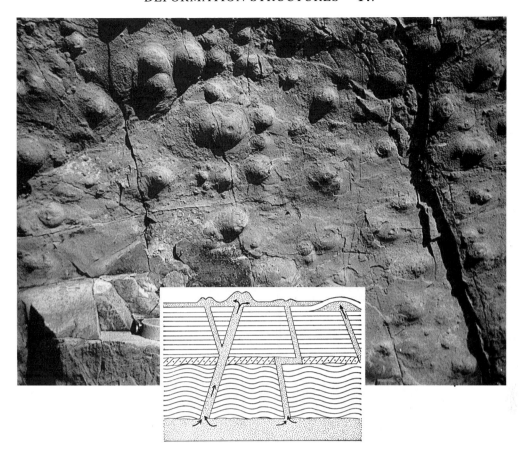

Plate 110
Injection structures: sand volcanoes

Sand volcanoes are moundlike structures whose diameter can vary from a few centimeters to several meters. They usually have a shallow central pit and less steep flanks than mud volcanoes. The last character obviously reflects the difference between the two sedimentary materials: sand is has no cohesion and has a smaller angle of repose than mud. Mud is so intimate a mixture of water and solid particles that it behaves as a fluid of its own, with its density and viscosity. Sand more rarely forms a single fluid phase with water; when it does, we say that it liquefies. Sand *liquefaction* requires particular circumstances: loose packing (which means hosting a lot of water in pores), fine grain size (increasing the surface to volume ratio, hence the surface in contact with water, and decreasing permeability), and some shaking (for example, by seismic waves transmitting stress to water). Liquefied sand, or *quick sand,* is squeezed out by the weight of overlying strata (overloading) and the internal pressure of pore water. Mud, in addition, can be pushed up by ascending gases.

The example illustrated here is found in Ancient turbidite deposits of the Hawick Rocks (Silurian of Scotland); identical Modern analogs have been described in a Modern tidal flat of the Arctic Canada by a sedimentologist endowed with a vivid imagination (he called them *monroes*).

Sand volcanoes can occur on both subaerial and subaqueous interfaces. Consequently, they are not environmental indicators.

Injection and fluid escape structures *can* be caused by earthquake shocks, but not necessarily so. It is not correct, therefore, to call them *seismites*, because this means assuming a univocal, deterministic link between cause and effect. Instead, one should make the hypothesis that this link exists on a case by case basis. This implies considering other, independent evidence in favor of seismic events. At the moment, there are no safe criteria to identify with certainty earthquake-produced structures in the fossil record. Sedimentary and other external mechanisms, acting alone, can account for fluidization, liquefaction and related phenomena, as will be exemplified by the illustrations that follow.

Seismite is a genetic term, and should not be employed for describing deformational structures (and less so, whole beds and depositional units). Only when the object has been discussed and interpreted, can a genetic term be *proposed*, being conscious that other interpretations are possible and that the preferred one is subject to confutation.

Photo: E. K. Walton 1970.

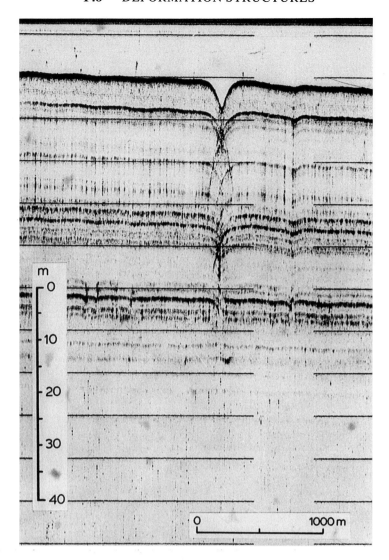

Plate 111
Injection structures: pockmark

Pockmarks are pits on the sea bed representing the superficial expression of large-scale gas vents. The example shown here is a seismic section recorded in the Adriatic Sea by a 3.5 kHz subbottom profiler. The geometry of the vent cannot be resolved in this record because of distortion of reflected waves ("hyperbolic echoes"), but is visible as a discontinuity (with downward inflections) in buried reflectors.

The deep beds from which the gas derives are not reached by acoustic waves, which have been absorbed by overlying layers.

Photo: P. V. Curzi and the Institute for Marine Geology C.N.R. 1992.

Plate 112
Dewatering structures

When sand liquefies, all preexisting structures are canceled. Injecting sand does not necessarily reach the surface, even though it tries to; it can form *blind* injection structures within buried beds (see inset to plate 110). These bodies are sheetlike, and are called sand dikes, or *sedimentary dikes,* if they cross the beds, or *sedimentary sills* if they intrude along bed surfaces. The terms derive from the igneous equivalents, caused by magma injections. Dikes can utilize previous fractures but in most cases are created by the fluid pressure itself (hydraulic fracturing). The "mother bed" of liquefied sand is thinned or reduced to discontinuous lenses; it can also disappear completely.

If liquefaction does not occur or aborts, only water escapes upwards through the porous sediments. It can do it gently, without displacing the grains, or with a violent, turbulent motion, in which case it entrains solid particles (elutriation) or deforms the sediment along its path. *Water escape* or *dewatering* structures will be the result. They are frequent in sands that accumulate quickly through catastrophic processes, as in the turbidite bed shown here, but can be also found in tractive deposits of shallow marine and alluvial environments (see color photo 6), especially in loosely packed levels. Dewatering structures are regarded as seismites by some geologists who ignore caution in labeling things (see caveat in commentary on plate 110).

The example reported here, from Eocene turbidites of Guipuzcoa Flysch near San Sebastian (Spain), shows that water expulsion was related to a depositional event. Blind deformations are here observed (they die out within the bed). In the lower, coarse portion, a faint lamination is visible; it is distorted by dish structures (see next two plates). In the overlying laminated sandstone, isolate cusplike structures occur; they pass upwards to attenuated undulations and eventually to a thick interval of plane-parallel laminae. In essence, water escape started as a massive phenomenon and faded out as more and more sediment was being deposited. At the beginning of deposition, the sand was more permeable, and the sedimentation quicker and more massive; higher up in the bed, both permeability and rate of sedimentation decreased.

Plate 113
Dewatering structures: dish

Plate 113 shows a detail of a vertical bed; to observe the structures in their natural position, rotate the page 90° counterclockwise.

In well-cemented sandstones (see also plate 112), *dish structure* looks like upcurled, partly intersecting dark lines. These lines represent enrichments of matrix components in the sand, made by selective deposition (laminae) or secondary concentration of fine particles entrained by ascending water (a sort of micro-sills). They formed semi-permeable membranes that resisted water pressure but could not prevent its escape, being thus deformed and disrupted. The formation of dishes should be progressive, from bottom up, but this cannot be easily demonstrated.

Dish structures can be used as way-up *indicators* in tilted and overtunred beds as exemplified by this sandstone unit in the Franciscan Complex cropping out in a California beach.

Plate 114
Dish-and-pillar structure

In weakly cemented sandstones, dish structure is emphasized by differential lithification. More cement was precipitated where the sediment was more permeable, in between matrix-rich films. Dishes are here associated with *pillar structures,* sort of small dikes cutting through bedding; some of them are short and blind, some others reach the bed top. Pillars are the better cemented portions of the bed and represented preferential routes of water escape during bed accumulation. Sand grains were probably lifted up by elutriation along them.

Summarizing, a tripartite zonation can be made in this bed: 1) the lower part is almost structureless; it probably underwent liquefaction and constituted the main source of escaped water; 2) in the intermediate portion, about 20 cm thick (see Jacob staff for scale), dish structure is predominant with some hints of pillars; 3) in the upper portion, dishes are subdued whereas pillars are standing out. The oblique ridge at midsection is a blind injection of larger size, sort of dike.

This is one more example of a "proximal" turbidite bed in the Marnoso-arenacea Formation near Sarsina, northern Apennines. It must be stressed, however, that dish and water escape structures are found also in shallow-water, littoral, and fluvial sands. They are *indicators* of process, not of environment. One of the conditions for their development, the poor, unstable packing of sand grains, can be attained in more than one way: rapid deposition by a catastrophic process is one of them; others can be provided by post-depositional disturbances, such as sediment churning by organisms, seismic shocks, fluctuations of pore pressure due to the passage of storm waves.

Marnoso-arenacea Formation, Umbro-Casentinese Road, Romagna.

Plate 115
X (bow-tie) and Y-shaped dewatering structures

In plate 5, a turbidite *megabed* has been presented in panoramic view; here are two close-ups of it showing plane-parallel lamination and peculiar dewatering structures, with a pencil for scale. It can be seen that laminae are not equally developed: they are fainter in some levels, well defined in others. Moreover, short "joints" are present in some laminasets, absent in others. They are not perfectly rectilinear, and part of them has a Y or X pattern. At first sight, these features, here described for the first time, look like tectonic structures, but one wonders why they do not cross the whole bed. This is rather puzzling. Upon closer inspection, most X- and Y-shaped "fractures" do not appear as fissures but as projecting, more cemented "ribs" (**B**).

The ribs can be interpreted more or less like pillars, i.e., as dewatering structures. This sounds reasonable, if you consider that each level with ribs rests upon a level with semiobliterated structures, well fitting the role of partly liquefied source bed for escaping water. Well-laminated levels had a relative rigidity because of tight grain packing, and were punctured by discrete injections.

The X-shaped structures, resembling bow ties, sandglasses, or a variety of Italian pasta, seem to be formed by a lower funnel where water converged, and an upper one where divergence and dispersion occurred.

Contessa Bed, Marnoso-arenacea Formation, northern Apennines.

Plate 116
Convolute lamination

This structure consists of wavy and contorted sets of laminae occupying a distinct interval in a sandstone/silt-stone bed or the bed as a whole. Their salient geometric features are: 1) the deformation does not involve base and top of the convolute interval; convolutions die out toward both boundaries, and are wholly *intrastratal* features (in contrast with slumpings); 2) the deformation style is entirely ductile, i.e., represented by folding of variable curvature, without any fracture: laminae remain continuous even when they are extremely deformed; 3) antiform folds (convex up curvature) are narrower than synforms, and their hinges have a smaller radius of curvature; 4) axial planes of folds are either vertical or dip upcurrent (convolutions mostly occur in current-laid deposits); 5) in some cases, the convolute laminasets are detached from underlying sediment along a *décollement* surface, corresponding to a level of weakness (plastic or semifluid material, such as matrix- or vegetal-rich sand).

The above-described features allow the structure to be used as a way-up *indicator* in tilted successions. From the textural point of view, the convoluted sediment consists of fine sand and silt; clay and vegetal particles can be present, adding cohesion and plasticity. Various proportions of these materials and water make the ductile behavior possible.

The origin of convolute laminae does not seem attributable to a single process. One possibility is that they are produced by water escape; this can be inferred by their frequent association with the structures just described in previous plates. The antiforms would represent escape routes while the synforms would simply be a "passive", complementary record of the deformation. According to another explanation, density contrasts *within* the laminated packet would put it in gravitational disequilibrium: if denser parts stay on top of less dense ones, the situation is unstable and the packet is solicited to overturn to restore stability. Descending parts (synforms) would be the active ones in this case.

A third interpretation regards the convolutions as a result of current drag. The current from which the sediment was just deposited would keep flowing over it, exerting a shear stress on its surface. Solid friction due to cohesion would drag the ductile laminae down to the depth where the sand reacts more rigidly. This mechanism seems applicable, however, only where the convolution planes are inclined and show a vergence in the current direction.

It is also possible that different processes act together or that each of them is responsible for specific cases. In any case, convolute laminae *are not caused* by gravity sliding, slumping or creeping, which means that they cannot be used as *indicators* of paleoslopes.

Marnoso-arenacea Formation, Santerno Valley, northern Apennines.

Plate 117
Some details of convolute laminae

Convolutions are markedly and geometrically complicated structures, as can be seen in these pictures. Picture **A** is a section of a sandstone bed in place, picture **B** of a fluvial pebble collected on the bed of a river eroding similar sandstones. However a packet of convoluted laminae is cut, the patterns we see is one of convoluted and wrinkled lines with various radii of curvature, similar to those obtained by stirring two immisicible liquids in a cup or jar. It reflects many tight folds made by soft sediment laminae slipping against one another. The fold axes have a limited lateral continuity, and are folded themselves; that is why linear trends are rare and short.

A: *Inoceramus beds, Cretaceous of Polish Carpathians;* **B:** *Marnoso-arenacea Formation, northern Appenines (scale in centimeters).* Photos: **A** *S. Dzulynski 1970;* **B** *V. Rossi 1992.*

Plate 118
Convolute and oversteepened foreset laminae

In this wholly laminated bed of fine sandstone, different types of lamination are superposed: parallel laminae, cross-laminae, and convoluted laminae. As an exercise in fine stratigraphy, you can try to delimit the various intervals by overlapping a transparent sheet.

Convolute laminae show a "disharmonic" curvature, with different radii, and a leftward vergence. In the central cross-laminated portion of the bed, you will note that some foresets are steeper than the angle of repose (which usually does not exceed 20–25° in these materials). The dip of both convolute and foreset laminae are consistent in indicating a paleocurrent flowing from right to left. The

shear, or drag force imparted by the current to the newly deposited (and, in part, still accumulating) sediment, was the mechanism responsible for the deformations. Remarkable is the fact that rippled sediment reacts with a mass behavior, when we know that free movement of individual sand grains is required to form ripples. Clearly, the sand acquired some cohesion soon after deposition and "froze" the cross laminae. The presence of some matrix can explain this phenomenon (remember that ripples in turbidites, crevasse deposits, etc. are not purely tractive but tractive-fallout structures).

Marnoso-arenacea Formation, northern Apennines.

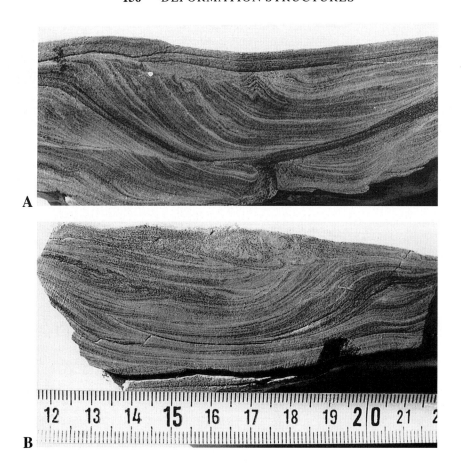

Plate 119
Current drag effects on foreset laminae

Some of the deformations illustrated in the previous plate are here shown in more detail. They occur in thin, cross-laminated turbidite beds, i.e., in traction-plus-fallout deposits. Laminae are emphasized by compositional and textural differences, the darker ones being rich in clay and organic (vegetal) matter (see also color photo 7). The grain size is in the coarse silt to fine sand range. Porosity and permeability are relatively low; some plasticity and cohesion is conferred to the sediment by finer and carbonaceous particles.

The paleocurrent (with opposite direction in the two specimens) first formed migrating ripples and cross-laminae, then deformed them in two ways: 1) by oversteepening the foresets, which remained anchored at the base but were dragged at the top; 2) by producing microfolding, or small-scale convolutions. Some microfolds were induced by gravity failure along the foreset slope (A). All such deformation occurred before the youngest laminae were deposited, as the upper picture demonstrates.

It is conceivable that, under the pressure of the current, *selective liquefaction* occurred: some sand laminae more prone to liquefaction, comprised between less permeable laminae, could yield to pressure and flow.

Marnoso-arenacea Formation, northern Apennines.

A

B

Plate 120
Current drag effects on cross-bedding

Oversteepening of foreset bedding was produced at a much larger scale in these examples, from the Miocene Valreas Sandstones, Rhône Valley, France. This unit was deposited in a shallow shelf sea, where subaqueous dunes or sandwaves were formed by tidal or storm currents. Along with oversteepened laminae, you can see here that the top of a thick foreset was hooked and overturned in the current direction (plate 120 **A**), and that master bedding surfaces, on which foresets lie, were also involved in deformation, which culminated in recumbent and cascade folding to the far left.   *Photos: G. Ori 1992.*

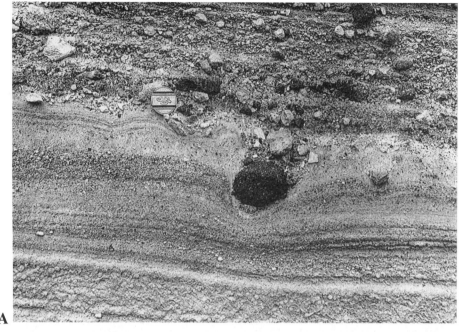

A

B

Plate 121
Deformation by impacting objects: bomb sags

Pieces of lava ejected by volcanoes (cinder) may spin and assume a spindlelike shape during their flight; a solid crust starts to envelop them. These *volcanic bombs,* upon landing on fine-grained, compressible materials like ash or mud, sink and leave more or less deep imprints, called bomb sags (**A**).

If the lava is still plastic or semifluid when it impacts, the sag is shallow and the bomb squeezes and deforms like a liquid, with a curious affinity to dung (**B**).

Quaternary pyroclastic deposits of Lipari, Eolian Islands (**A**) *and Procida* (**B**), *Tyrrhenian Sea.*

Plate 122
Load deformation: section view

The production of post-depositional structures is often stimulated by the load of sediment (or ice, lava, and other materials, including artificial ones) on buried surfaces, in particular on *interfaces* between sediments with different physical properties. It may happen that a material of higher density (e.g., sand) rests on a less dense material (e.g., mud). The intervening surface is potentially unstable, because of the inverted density stratification. The sand tends to sink and the mud to inject upwards. A convective-like motion starts, but the attempts of the two "fluids" to overturn the bedding and replace each other abort at a certain stage owing to frictional resistance. The deformation is thus frozen, as in the example shown in this picture. Its geometry depends, among other things, by differences in viscosity between sand and mud.

More commonly, mud is more viscous; the resulting profile is a lateral alternation of rounded down bulging lobes of sand and narrower, pointed crests of mud ("*flames*"), as is seen in convolute laminae. The profile would be inverted if mud had a lower viscosity than sand. Lobes and flames have a vergence in some cases, which indicates that some slippage occurred along an inclined

plane or (less probably in thick sand beds) that the buried interface felt the drag effect of a current.

The outcrop belongs to Messinian clastics of Marche region, central Apennines. Other examples of load structures are in color photos 3 and 59.

The elevated specific surface of its small particles gives a mud the property of cohesion, which increases with compaction and density. This implies that an applied stress can deform the mud and make it flow only when a critical, or *threshold* value is reached. Cohesion can be instantaneously lost if the mud has another property, *thixotropy;* when shaken, its gel structure is destroyed and it is transformed into a sol. In other terms, it liquefies. When the stress ceases, the sol spontaneously gelifies and stiffens. Both cohesion and thixotropy are related to the colloidal state, in which electrochemical forces are more important than gravity.

The mechanical disturbance that exceeds the cohesive resistance or annihilates cohesion can be provided by earthquakes, and consequently load structures could be also called seismites. For the reasons already discussed, however, this term should be avoided.

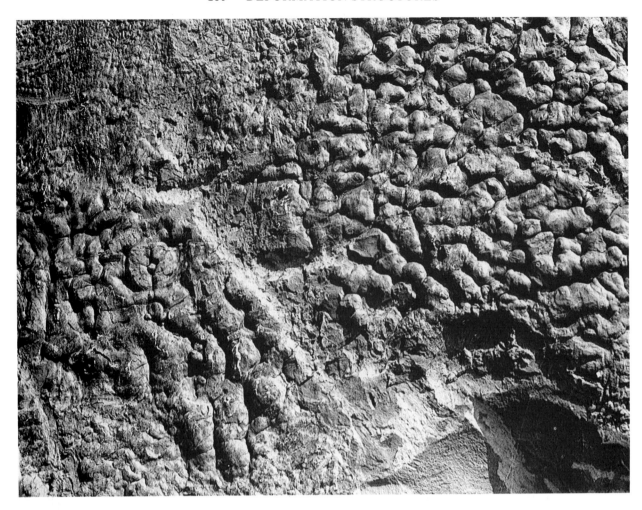

Plate 123
Load structures: plan view (load casts)

We can see here, on the base of a sandstone bed, that load structures are equidimensional in the horizontal plane; they are organized in polygonal cells as in typical convecting fluids (compare, for example, with the "granular" surface of the solar photosphere), the main difference being that density contrasts are not temperature-dependent in sediments. Some alignments of cells are also visible, giving the pattern a brainlike aspect. It is possible, in fact, that the load-induced polygons mask previous elongated structures made by a current (the deposit is a turbidite). I could also dare say that the supposed current went from the upper left to the lower right.

Pietraforte Formation (Cretaceous), northern Apennines; this is one of the deep-water siliciclastic "flysch" of the Mesozoic Tethys paleo-ocean.

Plate 124
Current marks deformed by loading

The rough base of this verticalized and overturned sandstone bed shows more clearly that load structures did not start on a smooth surface, but emphasized and overprinted previous current-produced structures. Some flute casts and one groove cast are well recognizable, and the paleocurrent direction can be determined. In other parts of the bed surface, current marks have been deformed or replaced by the brainlike convective pattern (see lower right).

Marnoso-arenacea Formation, northern Apennines.

A

B

Plate 125
"Squamiform" load casts (A) and load-casted ripples (B)

Load structures can assume these curious aspects at the base of thin, cross-laminated beds. The term "squamiform load casts" was invented in the Fifties, when most sole marks of turbidites were defined and classified. Structures like these had caught the eye of geologists years before, and received more fanciful names, like "dinosaur leather."[2] In this imaginative inclination, one could also say that they resemble the marks of paws.

The elongation of the load structures in the current direction, as can be checked by foreset laminae visible in section (**A**), seems to indicate that, also in this case, current marks were present on the bed surface before loading. It is possible that partial liquefaction of fine sand laminae contributed to the deformation.

In plate 125 **B,** brainlike sand pockets, thicker than the feeding bed, apparently grew by stacking of successive smaller lobes. The lobes were somewhat stuffed into the mud, with a component of lateral rotation. The interpretation is that the load was combined with current drag in forming these pockets as outgrowths of a pinch-and-swell structure (compare with plate 94). The troughs of ripples would have sunk more than usual and in successive stages, whence the term *load-casted ripples*. Barely visible foreset laminae indicate the paleocurrent direction.

A: *Marnoso-arenacea Formation, northern Apennines.* **B:** *Kliwa Sandstone, Polish Carpathians.*

Plate 126
Ball-and-pillow structure (pseudonodules)

Load-formed sand lobes, sinking into a soft muddy substratum, can detach from the "mother bed" or remain linked by a narrow neck or "umbilical cord." The resulting pockets and pillow-shaped bodies were called *pseudonodules* by European geologists, *ball-and-pillow structure* by Americans.

The structure can be reproduced experimentally by putting a sand layer on thixotropic mud and shaking the container, or by using two immiscible liquids of different density and viscosity (with the denser one obviously staying atop). The experiments demonstrate that load and inverted density gradients are not sufficient to generate the descending "drops," unless they are helped by liquefaction of both the substratum and the sand. If, in their descent, the sand drops meet an increasing resistance by stiffer mud, they can spread laterally; alternatively, they are flattened by subsequent overloading (with continuing sedimentation and deeper burial).

Buried horizons of thixotropic mud resist compaction because water expulsion is hampered; they remain in an *underconsolidated* state and represent horizons of weakness. Under a sloping bottom, such horizons are a cause of instability and can be involved (with overlying beds too) in slumping or debris flow phenomena. If balls and pillows were embedded in them, they would be found as *displaced* structures (the term *slump ball* was also used).

The picture, like those shown in plates 98 and 99, has by now a historical value as the outcrop has completely deteriorated (Marnoso-arenacea Formation, northern Apennines, road Fontanelice-Casola Valsenio). Another example of ball-and-pillow, a classical one reported in the atlas by Pettijohn and Potter,[3] is shown in color photo 8: the laminar flow of liquefied sand is there made visible by twirled decorations reminding classic capitals of ancient Greek temples.

Inset: modified from P. H. Kuenen 1958, Experiments in geology,
Trans. Geol. Soc. *(Glasgow), 23: 1-28.*

Plate 127
Enterolithic folds

The contorted and tightly folded layers of these pictures are composed of white, saccharoidal gypsum (calcium sulfate); we first met this variety of gypsum (*alabastrino*) in the evaporitic-anoxic deposits seen in plate 49. There, the beds were plane-parallel or only slightly undulated. Here, instead, they are deformed. The origin of this type of deformation has been hotly debated, and more than one explanation is possible, as was pointed out in the case of convolute lamination (plate 116).

When evaporitic environments were poorly known, *alabastrino* gypsum was thought to derive from the swelling of a previous sulfate, anhydrite, through hydration and recrystallization. Gypsum is a hydrated mineral, in fact, whereas anhydrite, as the name indicates, is anhydrous. Incorporation of water molecules in the crystal lattice implies an increase in volume. The secondary origin of *alabastrino* proved right but not the expansion theory. If the original aggregate of anhydrite crystals forms an open mush, when they are hydrated into gypsum the expansion can be absorbed by the pores and does not cause a swelling of the aggregate.

When inland and coastal *sabkhas* were explored (these are sedimentary environments barren of vegetation and encrusted by salts in arid areas), anhydrite aggregates were found just below the topographic surface, in the form of nodules (see plates 172–174), and laterally coalesced nodules forming bands, laminae or beds, often with enterolithic folds (**A**). Nodules with the same color and shape occur in the fossil record, but are made of secondary gypsum (or, sometimes, other minerals, like quartz). This makes clear that replacement of anhydrite by gypsum does not change the volume and the shape of nodules. The observed deformations, therefore, cannot be attributed to hydration and swelling; they are, instead, produced by the growth mechanism of nodules.

Anhydrite nodules grow in a host sediment but do not accept impurities inside (whence their bright white color); therefore, continued growth displaces and compresses the embedding sediment. If the expanding nodules are confined both vertically (by the weight of overlying sediments) and laterally, the only outlet left to them is lateral compression and folding.

This actualistic approach to nodular and enterolithic structures in evaporites explains very well the case of plate

128 **A,** where the gypsum is host in red continental deposits of Permian age (Bellerophon Formation of southern Alps). Its validity is more dubious when the structures are preferentially located around faults and zones of tectonic deformation (**B,** Messinian "serpentino" from a Sicilian mine or "solfara"); moreover, the gypsum is embedded in dark, bituminous marine marls like those shown in plate 49. These findings suggest the possibility that plastic gypsum mush or sulfate-rich solutions are injected under pressure by tectonic stresses, which could also cause tight folding (see plate 178).

Plate 128
Brecciated structure

A breccia can be viewed both as a deposit and as a kind of structure, or fabric; in the latter respect, it indicates an extreme deformation affecting rigid, brittle materials. A breccia is often a puzzling geological object as it can be originated by several processes, not only sedimentary ones. A volcanic or a tectonic breccia, for example, can be produced in at least twenty different ways, according to experts. I shall here enumerate the more common cases of *sedimentary* and *pyroclastic* breccias, and their causes. They are: 1) mass flows (fresh clasts of whatever origin are not rounded in matrix-supported transport); 2) fragmentation caused by sliding; 3) fall of single particles from rock cliffs (talus debris); 4) ballistic fall of pyroclastic products; 5) in place production of intraformational (mud) clasts by desiccation, expanding crystals or mechanical action of organisms; 6) collapse of solution (karst) cavities; 7) fluid injection (water, gas-charged mud) from below; 8) substrate liquefaction beneath cohesive or brittle beds, and, last but not least, 9) earthquake shaking (which again poses the problem of using the term seismite!).

As a consequence, a brecciated structure can represent many things and cannot be a reliable *indicator* of any in particular. In most cases, the interpretation of a breccia needs the help of independent evidence and cannot be given if the geological context is ignored or unknown.

This specific example shows elements of bituminous marl embedded in a carbonate matrix at the base of an evaporitic formation (Gessoso-solfifera, northern Apennines). Several processes, among those listed above, could have contributed to its origin, acting alone or in combination; namely, 1, 2, 5, 6, 7, 8, 9.

Plate 129
Deformations related to sliding and slumping

Landslides and subaqueous slides remove rock and sediment masses from slopes and accumulate them downhill. The movement is caused by the pull of gravity acting on the mass itself, in particular by the tangential (parallel to slope) or shear component of the mass weight. This force must overwhelm the resistance (shear strength) of the materials, which is function of their physical properties (cohesion, internal friction). By increasing the slope angle, the pulling force increases too, while the resistance does not vary. On the other hand, with a constant slope angle the resistance can be weakened by sediment liquefaction, excess pore water pressure, gas injection and other causes.

Underwater slides involve more or less compacted sediments and transfer them into deeper water. The volume of subaqueous slides is greater, on the average, than that of subaerial slides, the movement being favored by hydrostatic pressure that increases buoyancy and decreases friction.

In some cases, sliding masses are not deformed inside; they glide as rigid blocks on a slippery surface (*slip plane*) where deformation is concentrated. Before the failure, this surface is a horizon of greater weakness, such as an underconsolidated layer with a high pore pressure. More

commonly, sliding masses become disarticulated into a series of blocks, which move almost independently by sliding against each other along several slip surfaces. Eventually, deformation is diffused through the mass, as illustrated here; this is favored by the interbedding of materials of different "competence" (strength), represented in this outcrop by compacted sand and mud. Sand beds look like broken and folded strings (or the bolognese pasta "tagliatelle") with so-called *cascade folding* style. In mud beds, the deformation is more pervasive and less recognizable from a certain distance.

Is it possible to recognize the sense of movement in these folded structures? Sometimes it is, sometimes it isn't. It depends on the availability of certain information (see plates 131 and 134, for example). In the present case, the answer is no, because geometric data are insufficient: hinges and axial planes of folds dip in various directions, as do shear planes cutting the outcrop in blocks. In other terms, there is no preferred orientation of structures, which suggests that the movement of sliding masses has more degrees of freedom than tectonic movements occurring at deeper levels, which are more constrained by load and lateral confinement.

In conclusion, fossil slide structures are *indicators* of

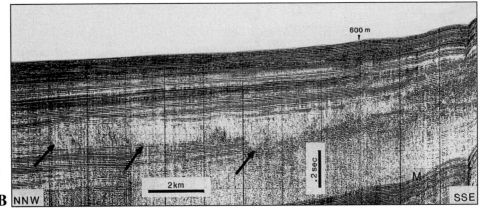

paleoslopes with some limitations; they witness to the former existence and instability of slopes, but not necessarily to their orientation.

Ancient slides are (improperly) called *intraformational* if they include intrabasinal materials, *extraformational* if composed by extrabasinal sediments or "exotic" rocks. Actually, what is "intra" or "extra" is not the process but the provenance of the materials. Intrabasinal sediments may have been originally deposited on either sloping or horizontal surfaces; in the latter case, tilting occurred after deposition, and can be inferred from the facies of slid beds. The turbidites shown in this picture, for example, were initially deposited on a flat deep sea plain.

Marnoso-arenacea Formation near Sarsina, northern Apennines.

Plate 130
Submarine slides: morphology and geometry

Two seismic sections across the Italian continental margin in the Tyrrhenian Sea document the aspect of a Modern (**A**) and an older (**B**) submarine slide. Record **A** was obtained by a 3.5 kHz Subbottom Profiler, record **B** by a 1 kJ Sparker source.

The Modern slide is indicated by the undulated topography of the sea bottom, which is also expressed in underlying reflectors. The undulations are related to the stepped profile of a body sliced by seaward dipping slip planes. The planes join a basal detachment surface (W). Almost horizontal beds lie below the slid mass, and an unconformity (U) separates them from clinoforms (or tilted beds) which represent the deepest levels penetrated by acoustic energy.

In **B,** the slide body appears as a structureless lens sandwiched between slightly dipping reflectors. The overlying ones have buried and "sealed" the slide. Their thickness (here expressed as a two-way travel time of acoustic waves: average sound velocity in Quaternary muddy sediments can be assumed as 1500 km/s) is considerable, suggesting that the slide is relatively old. The reflectors underlying the slide are parallel and continuous in the distal area (first two arrows from the left), becoming more irregular in proximity of the slope to the right.

The seaward tip of the slide lens displays surficial undulations and internal discontinuous, "chaotic" reflectors, while the rest is apparently structureless or "transparent". Such aspect is due to the disordered internal structure, which diffuses acoustic energy (compare with plate 2). This means that the deformation is intense and pervasive. The basal slip surface (see arrows) is more irregular than in section **A.** The upslope tip of the chaotic lens, to the right, joins an ill-resolved detachment zone, from which another slide of smaller size originated after the large one (see lens beneath the "600 m" spot). M is a multiple echo of superficial reflectors.

Photo: F. Trincardi, Institute for Marine Geology, C.N.R. 1992.

Plate 131
Ancient submarine slide: top unconformity

A sharply defined angular contact separates two parts of the same formation. The lithology is similar in the two parts (interbedded mudstones and sandstones, with prevailing mudstones), whereas the geometry of bedding is different. Beds are partly broken and folded in the lower part of the section, perfectly continuous and parallel in the upper half.

The original setting was a deep water basin with turbidite events punctuating hemipelagic sedimentation. A thicker turbidite stands out just above the unconformity; it represents a *megalayer,* i.e., a layer of great volume and basin wide extent, which can be used as a stratigraphic *marker* (compare with plate 5).

The deformed part of the section has a lateral continuity of some kilometers, and can be interpreted as an Ancient (Miocene) submarine slide, similar to the one of the previous seismic section. One would then expect a wavy surface at its top. The sharp contact means that the surface was smoothed by erosional and depositional events following the slide emplacement (see inset, plate 133).

The slide body is less intensely deformed than the one pictured in plate 129, with bedsets maintaining their parallelism (see right hand side). This suggests to apply the deformation model sketched in the inset, where up-slope dipping shear planes subdivide a mass accumulated at the toe of a slope. One of such surfaces cuts diagonally the outcrop (starting from the "V" shaped indentation to the right). If this interpretation is correct, the orientation of the paleoslope can be inferred.

Marnoso-arenacea Formation near Apecchio, northern Apennines.

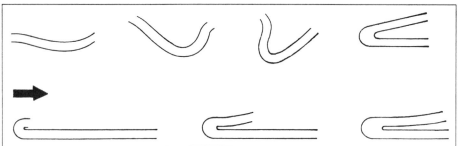

Plate 132
Folded beds in a paleoslide

We have already seen that folded structures tends to characterize slides made of heterolithic units, where the interbedding of more and less competent beds provides optimal conditions for the development of numerous surfaces of slippage. However, shear planes not only contribute to folding but also break the continuity of bedding. With deformation proceeding, disarticulated fold hinges are piled up in the style shown by this picture. Some features are worth pointing out in them. One is the parallelism of axial planes, dipping to the left and making a small angle with the horizontal plane. Possible ways of obtaining this *recumbent folding* are sketched in the inset

(the structures are also called *slump overfolds*).

Another characteristic is the thickening of sandstone beds in the hinges of folds, which suggests some flow of sand during deformation. Finally, the hinges are bracketed by banded mudstones in a sort of fork or wish-bone pattern (see lower left and upper right). Banding is a secondary character, acquired by viscous flowage.

The outcrop is part of a major slide mass, more than 100 m thick, interbedded in the Marnoso-arenacea Formation, northern Apennines, road connecting Palazzuolo sul Senio to Passo Sambuca (Florence).   Photo: G. Piacentini 1970.

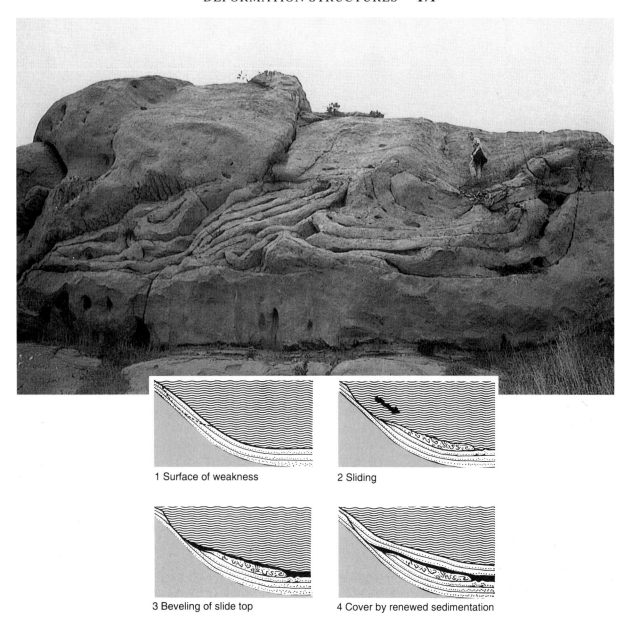

1 Surface of weakness

2 Sliding

3 Beveling of slide top

4 Cover by renewed sedimentation

Plate 133
Box-shaped folds in a sandy paleoslide

It is unusual to find fold-style deformation in thick and very thick sandstone beds. They commonly respond to stresses as rigid bodies (compare with next plate). The deformational motif is here a single recumbent fold with an angular hinge zone (box fold). *Disharmonic folding* can be noted: an outer shell, made by a huge sandstone bed, can be distinguished from an internal "stuffing" where thinner beds were almost rolled up. The outer shell outlines a single, major fold while the "stuffing" is made of several folds of smaller size and curvature.

The intercalation of mudstone beds is the key to explain this kind of folding. The plastic mud partings provided levels of weakness and allowed mutual movements of sand beds. The different thickness of sand layers made the rest: bed thickness constrains the radius of curvature of the hinge, imposing a relatively rigid box pattern to the thicker unit. Thinner beds could consequently be accommodated by tighter folding within the outer shell.

Plate 134
Rotated block in a paleoslide

The displaced character of this sandstone bedset is made obvious by its truncation and discordant dip with respect to underlying bedding. Moreover, its base is curve and cuts across the beds. A level of unbedded, dark mudstone separates the sandstone body from the substratum, made of mudstones with interbedded thin sandstones. The sandstone bedset is not deformed inside and clearly behaved as a rigid block during sliding. It also rotated and assumed an upslope dip, more or less like a person who lets his back slide on an armchair when taking a nap or watching TV.

Slided blocks are also named *olistoliths,* from a combination of two Greek words meaning "sliding stones". They can be found both as components (sort of giant particles or clasts) of slide and debris flow bodies or as isolated occurrences, not embedded in any "matrix". In the latter case, the blocks can either represent forerunners of a major slid mass or dispersed remnants of a completely fragmented body.

Marnoso-arenacea Formation, Santerno Valley, northern Apennines

---

A term often employed for a body made of chaotic and/or brecciated mudstones embedding rigid blocks is *olistostrome* (literally, "slided body"). The olistostrome concept was proposed by an Italian oil geologist (Flores) in the late fifties but not adequately defined. Neither the mode nor the speed of sliding was specified, or whether it should be regarded as a sedimentary or a tectonic process. Olistostrome has thus been used for tectonic mélanges and tectonic (friction) breccias on one side, for debris flow deposits on the other.

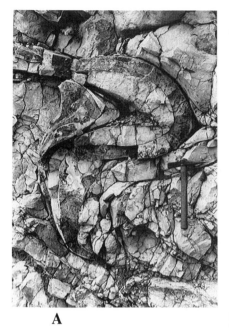 

A   **B**

Plate 135
Sedimentary and tectonic folding compared

Plate 135 **A** (enlarged on page 140) shows beds of lime-stone and chert (darker) folded in the shape of a question mark. The outcrop occurs in a slumped mass within a carbonate formation of Mesozoic age in southern Alps. In 135 **B,** beds of a lithologically similar unit are deformed into zig-zag or chevron folding, a typical tectonic structure due to compression.

Notice the differences in the geometry of folds: in **A,** hinges are rounded and disharmonic (with different radii of curvature), most beds being thicker there than on the flanks or limbs of the folds. In **B,** fold hinges are pointed, the folding more regular (repetition of the same basic unit) and harmonic (the beds keep their thickness throughout the fold profile). Furthermore, jointing is perpendicular to individual beds in **A,** thus showing all possible orientations, whereas it cuts through beds and is parallel to one flank of the fold in **B.** All these features indicate that the beds in **B** were deformed in more confined conditions and by a stronger, oriented stress of tectonic origin. In sliding and slumping, the deformation has more degrees of freedom.

The presence of chert in limestones deserves an annotation. Chert is crystalline silica (SiO_2) and is found in primary or secondary (diagenetic) beds. Silica occurs in the hard parts of small planktonic organisms (radiolaria, diatoms), in sponge spicules, etc. Siliceous remains form distinct beds or are components of carbonate oozes. Silica is highly mobile because of its solubility in seawater and tends to replace carbonate particles. The substitution is reciprocal: the tests of silica-secreting organisms are carbonatized. One can thus ascertain the migration of silica and its secondary deposition. The chert beds shown in this plate are indeed diagenetic. If they were already been present when sliding occurred, they would have been in a gel-like or plastic state. Lithified chert is crystalline and hard; it would not fold but break.

Alternatively, silica migration might have occurred after sliding, the sliding itself providing routes for it (joints and fractures). Late diagenesis, in fact, cannot use sediment pores; deep burial and compaction, or cementation, have eliminated them. Permeability, necessary for the circulation of chemical solutions and reactions, can only be created by fracturing.

Pore reduction with depth, and a consequent increase in packing and density of sediments, has an important bearing on sliding, too, because it increases internal friction and hence shear strength and stability. Sliding is thus a relatively shallow phenomenon if referred to burial (not water) depth.

Photos: from A. Castellarin 1964, Geologia della zona di Tremosine e Tignale (Lago di Garda). Giornale di Geologia (Bologna), 32: 291–346.

Plate 136
Details of a paleoslide: snapped bed

A sandstone bed, belonging to a relatively undeformed set, failed under stress. The two facing ends, at break point, were rolled up, with a clockwise rotation, which can be explained as an effect of shear, or drag: the upper beds moved to the right, the lower ones remained in place or moved less. Tension accumulated in the beds, until one or more of them (we have a limited view here) broke and released it. Such snaps in subsurface sediments and rocks send signals to the surface through seismic waves, and this is one of the cases where the term *seismite* seems more appropriate.

As an alternative explanation, one can imagine that the hooked bed was broken by an intrusion of mud from below (where the hammer is placed), a process known as *diapirism* and related to overloading and density effects. If this were the case, the hooks should be symmetrical about a vertical plane, as they would have rotated in opposite ways.

For stratigraphic unit and location, see plate 132.

A   B

Plate 137
Details of paleoslides: outcrop vs. core

Plate 137 **A** shows another detail of plate 132, this time from its uppermost part. A kind of crevasse was opened in the slide body (by the pulling apart of two blocks?), and fluid or plastic mud filled it. The sea bed was not much above. The crevasse stopped against a harder layer (bottom of picture).

The disturbed sediment shown in plate 137 **B** is fine-grained and was cored in an underwater slide body similar to that seen in plate 130. Small pebbles are embedded in it, which suggests mixing of different materials.

Apart from deformations induced by the coring process itself, no observations can be made on the style and geometry of disturbed structures. The lighter patch to the right, for example, could represent either a mud chunk or the end of a contorted bed. One can only say is that bedding is disturbed or chaotic. An important indication comes, however, from the pebbles; unless they were dropped by melting ice or other floating objects, their presence in mud can be explained by mixing of marine

sediment with beach or river derived gravel caused by a mass transport. A slide in a coastal zone could have evolved into a debris flow by incorporating water and disaggregating. In conclusion, the cored sediment may represent the *indirect* record of a slide.

Some decades ago, before the role of gravity in sedimentation was fully appreciated, most debris flow deposits were interpreted as glacial or periglacial sediments. A muddy sediment containing scattered clasts was invariably attributed to glacial transport and called a *till*, or a *tillite* if lithified. Subsequently, the terms *diamicton (diamictite)* and *pebbly mud (stone)* (when the clasts are rounded) were introduced for purely descriptive purposes. They do not imply any particular mechanism of origin, even though pebbly mudstones are commonly regarded as products of mass flows.

Photo: B Institute for Marine Geology C.N.R. 1970.

Plate 138
Paleoslides: detachment scars

Subaerial and subaqueous slides have similar morphologies, in which three parts are distinguished: the zone of detachment, where a morphological scar is left by the moving away mass, the zone of translation, or slippage, and the zone of accumulation. The intermediate zone can be short and virtually coincide with the last one. Extensional stresses dominate behind the detaching mass, compressional ones in front of it (see drawing in plate 131). The examples of slide deformation we have seen in previous illustrations concern, with the exception of the upslope portion seen in plate 130 **A,** the accumulation zone. To conclude this section, I examine now one of the few published examples of possible *slide scars* in the fossil record.[4]

Curiously enough, the outcrop is itself a Recent slide scar; in other words, Recent slides have made possible to observe an older (Miocene) analog. The geometry of the section is similar to a kind of giant trough cross-bedding, or to the fill of stacked, partially shifting channels. What

does not fit with these interpretations, however, is the fact that sediments are not coarse-grained; thereby, they do not represent deposits of high-energy flows. Neither are lag pavements present to signal possible bypass zones of channels. The rock is a thin-bedded, hemipelagic marlstone, where bedding has partly been obliterated by the activity of benthic organisms (bioturbation). A modest to moderate paleodepth can be inferred from the fauna.

It is thus reasonable to assume that the "erosional" surfaces are scars left by slides on the upper part of a paleoslope; the removed masses should be found in a deeper and more distal part of the basin (accumulation zones are recognized by an anomalous thickness of sediments, even if the slides themselves are not visible).

The strata filling and draping the scars make *onlap* contacts with them.

S.Agata Fossili Formation, Tertiary Piedmont Basin, Italy.

Plate 139
Close-up of slide scars

A detailed view of the previous outcrop emphasizes the abrupt angular contact along one of the scars. The unconformable surface is marked by a more pronounced cementation of the mudstone (another one is visible near the top). It is possible that this lithification is not an after-burial diagenetic effect, but occurred when the sediment was exposed on the sea bottom, during a period of very slow or absent sedimentation. In other terms, it could be the initial stage of a *hardground* (see plate 171), a form of submarine encrustation.

Although the distinction between a slide and an erosional episode is important, their effects are the same under the viewpoint of the stratigraphic record. Geometrically, they produce unconformable or discordant surfaces in the source areas, and "excess" thickness of sediments in accumulation areas and depocenters; temporally, they leave stratigraphic gaps, or *hiatuses*. For this reason, sliding and erosion could be encompassed by the term *denudation*.

Photo: P. Clari 1992.

ENDNOTES

1. R. L. Bates and J. A. Jackson, eds. 1987. *Glossary of Geology.* 3d ed. Alexandria, Va.: American Geological Institute.

2. G. H. Chadwick. 1943. Ordovician "dinosaur leather" marking (exhibit). *Bull. Geol. Soc. Amer.* 59: 1315.

3. F. J. Pettijohn and P. E. Potter. 1964. *Sand and Sandstone.* New York: Springer Verlag.

4. P. Clari and G. Ghibaudo. 1979. Multiple slump scars in the Tortonian-type area (Piedmont Basin, northwestern Italy). *Sedimentology* 36: 719–730.

CHAPTER 7
Biogenic Structures

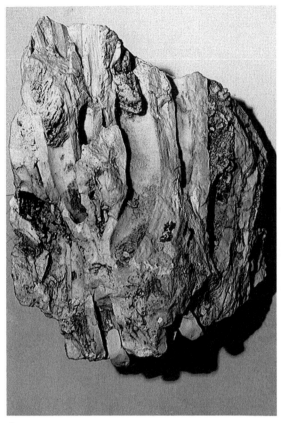

A   **B**

The presence of life, in its multifarious aspects, has an important bearing on both the origin and the modification of sediments. The ways that life and sediments interact are highly diversified, and the same is true for the morphology and the scale of their products. It is obviously impossible to illustrate them all, and a selection is necessary. Moreover, a sedimentologist cannot describe and understand structures related to life or biological processes (*biogenic structures*) without the help of other experts (biologists, ecologists, paleontologists).

In this chapter, some structures are presented that exemplify relatively common *categories* of biological activity related to sedimentation and, above all, provide useful and often valuable information on sedimentary processes, mechanisms, and environments. Among biogenic structures, in other terms, sedimentological *indicators* will be emphasized.

I will start with so-called *constructive* features, which contribute to sediment accumulation in some way or the other. After that, I will examine *destructive* and *biodeformative* structures, resulting from organic activities that disturb the sediment and obliterate, in part or entirely, its depositional structures.

Plate 140
Reef-building organisms in living position

Rudists, seen here in plate 140 **A** and full relief in **B,** are extinguished bivalves that had, more or less, the same ecological niche and function of Modern corals. They took root on a muddy or detritus-covered substrate and lived in crowded assemblages or colonies. When the organisms died, new ones grew on them pushing the surface of the colony upwards. Such colonies can be made of animals, algae or both; the organisms secrete calcium carbonate to build their skeletons and to bind each other in a rigid framework capable of resisting the mechanical disturbance of waves and currents. This sort of "pre-lithified" carbonate mass is a reef. A looser aggregation of sedentary organisms, preserved in their living positions but not cemented together, is an organic *bank*.

Organic reefs and banks grow vertically up to sea level, where further growth is prevented by subaerial exposure and mechanical breakdown; it can go on, however, to compensate for subsidence.

Reef-building animals of different *taxa* show a similarity of forms due to adaptation to a similar environment (a phenomenon known as convergence, in evolutionary terms). In a reef community, there is a preference for conical shapes, as happens here, and this is a useful way-up indicator (cone apexes point downwards).

A fragment of rudist lies horizontally in picture **B**; it was probably torn by storm waves.

The samples are conserved in the collection of the Department of Geological Sciences, Bologna University, and come from Cretaceous limestones of Carso (NE Italy) and Murge (SE Italy).   Photo: G. Piacentini 1970.

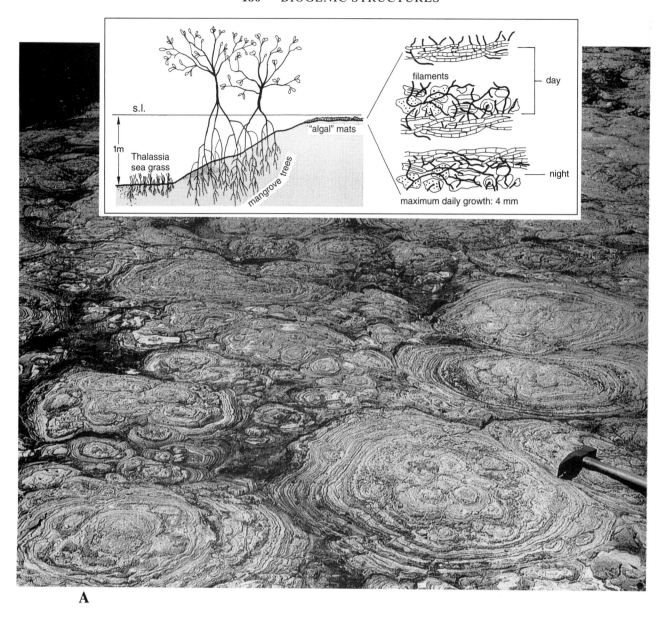

A

Plate 141
Stromatolitic structure

Living forms, especially vegetal, can contribute to sedimentation by acting as traps for sedimentary particles; this usually happens in shallow water, intertidal and swamp zones. Among the examples sketched in the inset, *algal mats* have a fly-paper effect on passing sediment; they capture it with their mucilaginous filaments. Actually, they should be called *bacterial mats,* as the constituent micro-organisms, which were regarded as unicellular algae, have been reclassified as cyanobacteria. Cyanobacteria represent one of the oldest form of life on Earth; the diffusion of their mats is restricted today by grazing animals, but was widespread during the early history of our planet. The bacteria rarely fossilize, but their mats do, and build a structure that is called *stromatolite.* It must be clear that a stromatolite is not, strictly speaking, a fossil, which means the mineralized remain of a living

organism, but a biogenic structure, i.e., the fossil evidence of the past activity of organisms.

A stromatolite is a lithified bacterial mat; lithification is due to recrystallization of fine carbonate matrix and precipitation of cement in pores. Most stromatolites are part of carbonate calcareous and dolomitic formations, where they have fixed enormous quantities of calcium carbonate during the first three billion years of Earth history. In so doing, bacteria have sequestered CO_2 from the atmosphere, thus contributing not only to sedimentation but also to a decreasing of the greenhouse effect and of the Earth's temperature. In other terms, the proliferation of primeval forms of life populated the planet in a way that favored the further development of life by keeping the temperature lower than the boiling point of water. Life seems thus to have set up a global self-regulated

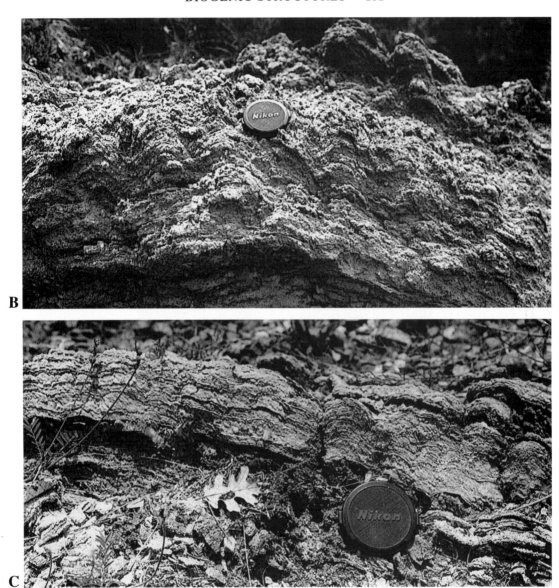

B

C

system apt to its survival. Today, the occurrence of bacterial mats is limited to the carbonate environments of the intertropical zone (Bahamas, Persian Gulf, Australia, etc.). Except in Shark Bay, Australia, they form no longer domal (plate 141 **A**) and columnar structures as in primeval seas, where grazing animals were lacking.

Stromatolites are less frequently fossilized in noncarbonate rocks; we see here two examples (**B** and **C**) of stromatolitic gypsum. In this case, the bacterial mats trapped a peculiar type of detritus, made of small gypsum crystals or fragments of larger crystals. These crystals of salt precipitated from evaporitic brines, to be later on resuspended by waves and storms. It has been observed in Modern environments that gypsum crystals can also precipitate within the mats, as bacteria can tolerate a high salinity. Crystals form in this case the seeds for cementing the mat (see plate 142).

As is obvious from these images, stromatolites are laminated structures; laminae are commonly wavy and crisped. Take care not to confuse their geometry with that of tractive (ripples) or convolute lamination. An indicative characteristic is the nonuniform curvature, with rounded convexities and acute, pointed concavities (in ripples and convolutions, it is just the opposite). Stromatolitic laminae reveal that the accretion of the mat is not continuous; there are, first, daily pauses (bacteria are photosynthetic organisms and interrupt their activity during the night: see inset), then seasonal or sporadic erosional events.

A: *Paleozoic carbonate of Petrified Sea Gardens, N.Y.;* **B** *and* **C:** *Messinian gypsum beds of maritime Tuscany, Italy.*

Plate 142
Bacterial mat embedded in gypsum crystals

This polished section shows an aggregate of large gypsum crystals (the variety called selenite). A graduated scale in millimeters is shown on the left side of the picture. In the upper part, the crystals are in place and form a sort of palisade, which rests on a substrate of detrital gypsum. They have a swallow-tail, or spear-head shape, deriving from their upward growing and twinned character. This orientation provides a way-up criterion, particularly useful in poorly bedded sequences (crystals converge downwards, according to the so-called *Mottura rule* applied in old sulfur mines of Sicily). Other examples of selenitic gypsum can be seen in plate 167 and color photo 30.

The horizontal lamination that is visible in the upper crystals is biogenic (which means "of biological origin"; therefore, one should not say "of biogenic origin"). They consist of superposed felts of tiny bacterial filaments (enlarged in plate 143). Bacterial mats and selenite crystals grew more or less simultaneously on a muddy bottom at the margin of a shallow-water body. The two processes were competitive, with both mat and crystals growing upward. The selenite incorporated the mat, which could survive only if the crystals did not outpace it. From time to time, violent storms broke the crystals and mats, accumulating layers of gypsum breccia and gypsum sand; bacterial mats formed again on them, as well as selenite crystals.

Messinian Vena del Gesso, northern Apennines.

Photo: P. Ferrieri 1976.

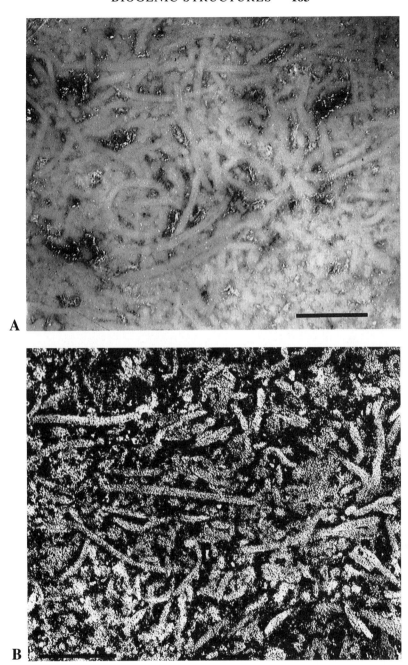

Plate 143
Microstructure of stromatolites

The two pictures illustrate the microscopical aspect of bacterial mats. The organic matter of the filaments is usually decomposed and does not fossilize. It may happen that the filaments are coated by a sheath of carbonate, which allows their preservation. With the addition of carbonatic or gypsum cement in the pores, the fabric, or microstructure of the mat can also be maintained. One can thus observe it in the most minute details either on a polished section (**A:** the bar is 1 mm long) or in a thin section (**B**). The difference between the two images derives from the fact that light is reflected and transmitted, respectively (the thin section is a slice of rock, so thin as to be transparent; the picture is a negative print). The microstructure of the mat resembles that of spaghetti, with some organic matter and iron sulfide in the "sauce."

Same formation and provenance as plate 142.

Photo: P. Ferrieri 1976.

Plate 144
Organic coatings on grains: rhodoliths

Stromatolitic laminae can coat also mobile objects, such as shells and pebbles rolled or suspended by water turbulence; they will then grow concentrically around a core and take the name of *oncoliths* or oncolitic structures. If the oncolith is being continually moved, the coating can surround it entirely; if, on the other hand, it sticks to the bottom, its growth can continue only on the upper, free surface.

Oncoliths look like the objects shown in these pictures, which by the way are not oncolites but *rhodoliths* (it cannot be written rhodolite as this is the name of a mineral). The coating is here made not by mucilaginous filaments and trapped sediment, but by thin crusts of calcium carbonate secreted by coralline red algae (*Melobesiae* or *Rhodophyceae*), which are benthic and need a firm substrate. Rhodoliths are real (body) fossils and real algal structures. The carbonate fixed in their tissues is micro-crystalline (micrite); it looks white in polished sections like those of the pictures, dark (opaque) in thin sections.

Both oncolites and rhodoliths form in shallow, agitated waters; in this respect, they can be used as *indicators*. More specifically, each of these structures characterizes different sub environments and ecotopes within carbonate depositional environments, from extensive platforms to more restricted banks, reefs, shoals, etc.

There are other types of coated and encrusted particles, whose envelopes are purely chemical precipitates. They are included under the heading "chemical and diagenetic" structures in this book, but could as well be classified, together with biogenic types, as *concretionary structures*. This is another example of the limits and subjectivity of every scheme of classification.

Miocene and Pliocene organic limestone (ancient fossil banks), northern Apennines.

Plate 145
Diffuse bioturbation and mottling

We have seen that many *sedentary* benthic organisms, both plants and animals, are sediment builders: they either trap passing solid particles or extract ions from water and build up carbonate accumulations. On the other hand, *mobile* benthos, represented by animals living on or under the sedimentary interface, disturbs and modifies already existent sediment. Part of its activity occurs above the bottom, and leaves *interfacial,* or superficial traces; another part consists in penetrating and crossing the sediment mass, which leads to *internal* traces. On the one hand, therefore, we have, broadly speaking and simplifying things, *grazing* and *crawling* organisms, on the other tunneling and *burrowing* organisms.

Burrows are distinct animal traces, whose preservation depends on their stuffing with organic matter (excretions) and sediment. They are sedimentary structures of their own, but at the same time, like an eraser, cancel previous structures along their pathway. As burrows grow in number, the degree of obliteration of primary structures and of sediment churning increases; the end product is an entirely homogenized, structureless bed or, as in the case illustrated here, a *mottled* bed. Mottling can be overprinted by secondary (diagenetic) effects, consisting for example in the selective precipitation of chemical cement, which gives the sediment a *nodular* or encrusted aspect. This depends on the fact that porosity and permeability are not uniformly distributed in the sediment, and vary from place to place.

Homogenized and mottled sediments are 100% *bioturbated.* Single animal traces cannot be identified, but there is nonetheless an important indication concerning the relative rate of sedimentation. In fact, rapidly accumulating sediments are rapidly buried as well; taking a relativistic stand, we say that they cross the depositional interface at a high velocity. This means that the sediments remain exposed to biological activity for a short time only, as the animals cannot penetrate beyond a critical depth (they need oxygen, among other things). Bioturbation will thus be minimal or absent when the rate of sedimentation is high. When and where sedimentation is slower, burrowers rework the sediment for a longer time and the extent of bioturbation increases. In conclusion, the amount of bioturbation is a rough, relative *indicator* of the deposition rate; fully bioturbated levels, in particular, reveal episodes of slow or interrupted sedimentation. For example, marine "blue" clays of Plio-Quaternary age filling part of the Po Basin and onlapping the margin of the Apennines, accumulated at a rate of about 1 mm/year, which is quite a high rate. At some levels, mottled sand-clay mixtures like those shown by the picture point out pauses in sedimentary supply. Some shells are scattered in the sediment, but burrowing organisms do not necessarily leave their remains; most of them are soft-bodied animals.

A   **B**

Plate 146
Diffuse bioturbation and burrow traces

The outcrop shown in plate 146 **A** is made of clayey-silty sandstone, whose bedding has been almost entirely obliterated by burrowing organisms: only "ghosts" of the original beds are visible (as in plates 20–22). The bioturbation was pervasive and led to *homogenization,* or biological amalgamation of distinct lithotypes. Notwithstanding that, individual structures can be seen in both the lower and upper part of the section. They are called, in general, *trace fossils (*or *ichnofossils*) to distinguish them from body fossils. Various types of traces can be recognized in Modern and Ancient sediments; they are classified by diverse criteria, such as the organism ecology and physiology, the trace morphology, its location and orientation with respect to bedding planes. Many traces were mistaken for real fossils, and received generic and specific names according to Linnean rules of paleontological taxonomy.

Coming back to picture **A,** two types of burrows occur in these "dirty" sands, which were deposited in a littoral (shoreface) environment of Pliocene age (compare with plate 21). Lower burrows are outlined by a black coating of manganese oxides; their shape varies depending on the orientation with regard to the section. The upper burrows

stand out because of cementation; they are smaller and more cylindrical, and were clearly made by a different type of animal. Curiously, they resemble "macaroni" pasta, which induced somebody to propose the name *Macaronichnus* for very similar structures found in Miocene sediments of northern California.

Pliocene Intra-apenninic Basin, Zena Valley, northern Apennines.

Plate 146 **B** shows, for comparison, a two-layered core, with burrow density increasing near the top of the light layer. What is the meaning of this variation? The white-black interface is probably the top of a rapidly accumulating bed (no traces are present in its lower part). It remained exposed for a while and was populated by benthic organisms that reworked the topmost part of the previous deposit. Penetration was less easy as soon as the burrowers approached the critical depth, where traces become scanty.

The darker mud at the top of the core accumulated more slowly, and was bioturbated throughout.

Plate 147
Vegetal (plant) trace fossils

Specific types of trace fossils provide information on the sedimentary paleoenvironment: vertebrate tracks or marks of plant roots, for example, will indicate a continental or coastal environment. In marine sediments, certain types or assemblages of traces characterize distinct depths and can be used as *paleobathymetric indicators*. Caution is needed, however, in paleoenvironmental reconstruction: do not forget that biogenic structures represent *modification* processes, which can occur in the same environment of deposition or after a change of conditions. A marine sediment can be bioturbated after emergence caused by tectonic uplift or sea level fall; the traces will then be made by continental organisms. On the other hand, a transgressed continental deposit will be overprinted with traces of marine organisms.

The picture shows a finely laminated lacustrine sediment crossed by subvertical, upward fanning traces left by the hollow stalks and foliage of a small palm tree (similar plants are rooted on the Modern soil: a glimpse can be caught in the background). Note the meniscus-shaped fill of the central hollow. The trace is a *passive* one, in this case: the plant left its mold on the sediment that buried it. Root traces are, instead, active, because this part of the plant disturbs the sediment (see color photo 6). Both active and passive traces, being preserved in place, are anyway indicative of environmental conditions.

*Rhythmic seasonal deposits (*varves*) of Oued Saura, Grand Erg Occidental, Algerian Sahara.   Photo: G. G. Ori 1992.*

Plate 148
Birdfoot casts

The base of a sandstone bed preserves the molds of tracks left by the feet of palmiped birds roaming the smooth muddy bottom of a coastal pond or a tidal flat. Some shallow elongated ridges hint at molds of mud cracks. The structures are obvious *indicators* of emergence. They occur on Cenozoic deposits of the Pyrenees, consisting of interbedded sandstones and mudstones that look like many "classical" turbiditic formations. And so in the late fifties they were interpreted as turbidites by the first zealots of the new paradigm of turbidity currents: flute and groove casts on the bed soles (not shown in our example) were carefully observed, but the birdfoot casts escaped attention. This caused a great controversy among sedimentologists, because opponents to the new idea of turbidity currents, mostly represented by some French academics, took advantage of this major turbidite mistake to slander turbidite fans. In fact, because of these famous *pattes d'oiseaux* (and some culture problems), the turbidite concept did not became fully accepted in France until ten or fifteen years later than elsewhere.

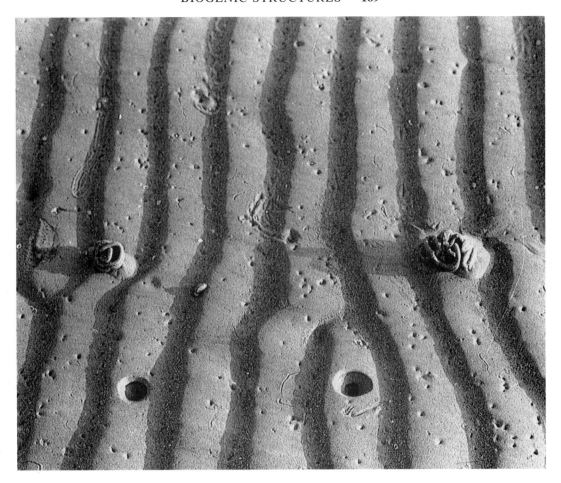

Plate 149
Vertical structures in shallow water: dwelling traces

The rippled sand of a beach at low tide is punched by many small holes and two larger ones; a heap of sand resembling squeezed toothpaste lies beside each larger hole. The holes are the entrances of the "homes" of a cylindrical worm named *Arenicola* (inhabitant of the sand), while the "toothpaste" consists of sand mixed with excrements heaped over the exits. By joining each hole with the respective heap, you get the width of a U-shaped trace that extends downwards for some decimeters. The living worm is still there.

U-shaped and similar traces (Y- or I-shaped), which develop mostly in the vertical dimension, are typical of nearshore environments with a sandy bottom, subject to moderately to highly agitated water. Not only worms, but also mollusks, crustaceans and other invertebrate animals make these *dwelling traces*. They live in the sediment because the interface is too disturbed by water motion; on the other hand, being filter feeders, they appreciate agitated waters because of their cleanness, which prevents filtering apparatuses from being clogged. Upon the organism's death, the traces are filled by sand and can fossilize. They thus become *indicators* of the littoral-intertidal zone in Ancient sediments.

Plate 150
Vertical structures in shallow water: escape traces

Bivalves, belonging to the species *Mya arenaria,* are here found in living position within nearshore sands of the North Sea coast. Below each shell, a vertical trace can be seen, which crosses and truncates depositional laminations. The filling sand is laminated too, but the laminae are biogenic in this case: they are concave-up and meniscus-shaped. The organism makes them by stuffing and compacting sand at the bottom of its living place. In so doing, it elevates its position. This means that the depositional surface is accreting, because each filter-feeding animal has its optimal depth of burial and tends to keep it constant. It is thus obliged to follow the ups and downs of the bottom caused by sedimentation and erosion, respectively. We know that the bottom position is not stable in the submerged part of beaches because of seasonal changes and exceptional events.

Sedimentation can be particularly rapid in the wake of a storm event; the organism trying to keep pace with it leaves an *escape trace,* or burrow; if it succeeds, the trace will reach the new interface, otherwise it will remain buried (blind). The meniscus-like traces, called *spreiten structure,* indicate the vertical movement of the animal. *Photo: H. E. Reineck 1970.*

Plate 151
Callianassa **burrows**

The outcrop shows more and less cemented beds of sand in Pliocene littoral deposits of Marche Apennines. Differential cementation emphasizes biogenic traces, which have both vertical and horizontal orientations, and locally cross each other producing irregular, web- like masses.

The burrows reflect the activity of a small crustacean, called *Callianassa,* which explores the sediment both for food and shelter. This trace is another indicator of near-shore environments.

A

B

Plate 152
Crawling (feeding) traces

We now pass to biogenic structures characterizing sediments deposited in deeper water, i.e., below wave base. These sediments basically consist of mud slowly accumulating by normal fallout of particles from the surface, or mud and sand emplaced by catastrophic processes (storms, turbidity currents). The organisms living there obviously like quiet water and can tolerate mud in suspension; they also get their food by ingesting mud and absorbing organic substances contained in it, which qualifies them as mud-eaters or *limivorous*. Most mud-eaters are soft-bodied and do not fossilize. Their prevailing movements are horizontal, vertical displacement being limited by the availability of oxygen (its rate of consumption can exceed the rate of supply in sediments of low permeability). Consequently, reptating animals (like the Modern gastropod *Bullia* in picture **B**) leave mostly surface traces, i.e., trails.

A sporadic event of high energy can bring in sand, whose deposition buries the traces and preserves them as molds. We thus find the structure as a sole mark (cast) in turbidite and tempestite beds. The meandering trace of picture **A** is an example; it was found in the Cretaceous Pietraforte Formation, a "flysch" of northern Apennines. Smaller circular marks, scattered on the bed surface, represent sections of vertical burrows.

Photo: G. Piacentini 1970.

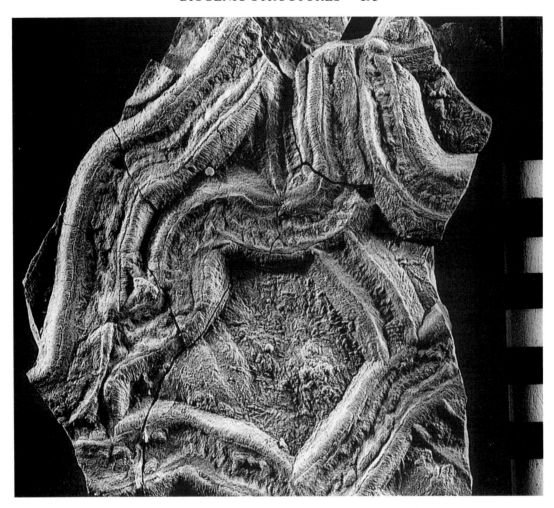

Plate 153
Crossing trails

These trails were molded by silt, whose fine grain size allows preservation of the slightest detail (the scale on the right is graduated in centimeters). In comparison with the previous example, this trace is characterized by crossings and overlaps. The morphological differences depend on the exploratory strategy of the animal. Most reptation traces, in fact, are made in search of food (*feeding traces*). Next plates document more cases in the same respect. This one shows the base of a siltstone bed deriving from varicolored deep-water shales with intercalations of thin-bedded turbidites of Oligocene age.

Northern Apennines (Marecchia Valley, Ligurian complex).
Photo: P. Ferrieri 1992.

Plate 154
Echinoid traces in section

This is a replica (peel) of a sandy sediment extracted from a shallow sea bottom with a box corer. The sample has been vertically sectioned, and shows a peculiar type of onion-shaped trace produced by the lateral movement of an echinoid (sea urchin). The sense of displacement of the animal was from right to left, and its body left a partial imprint at the end. Similar structures are partly visible near the upper left corner of the picture. They look like horizontal *spreiten* (turn the page 90° clockwise and compare with plate 150), and are originated by a similar "stuffing" mechanism.

The visibility of the biogenic structure is enhanced by its contrast with the primary, tractive lamination of the sand, indicating a shallow-water marine environment (above wave base). Echinoids, however, move the same way in resedimented deep-water muds and sands; the degree of preservation of their traces can be weaker there. Being found at various depths, this kind of trace is not a reliable indicator of bathymetry.   *Photo: H. E. Reineck 1970.*

Plate 155
"Problematic" traces: *Zoophycos*

Even if you do not know exactly what kind of animal produced a certain trace, you can interpret its morphology by analogy with Modern counterparts (see plate 152). Many trace fossils, however, were made by extinct organisms, and have no analogs in Modern sediments. For this reason, they were grouped in the category of *Problematica,* or problematic traces. This stimulated ichnologists to thoroughly analyze the morphology of traces to infer the kind of activity or of organism that could possibly generate them.

Zoophycos is one of such forms. It is not purely superficial but three-dimensional, with a helicoidal structure winding around a vertical axis. On the upper bed surface, it appears like an umbrella or a parachute with arcuate or spiraled ribs. In between the ribs, meniscus-shaped wrinkles are more or less evident. *Zoophycos* is typically found in hemipelagic marls and mudstones (the example shown here) or in limestones (see color photo 24). It is one of the more complicated traces and reveals a systematic and spatially ordered strategy for exploring the sediment as a source of food. Every cubic centimeter was exploited within a certain area and a certain sediment thickness.

In terms of water depth, this trace characterizes a wide range, which is intermediate between the shallow-water zone with predominant vertical traces and the deepest recesses, where mostly superficial tracks occur.

Miocene marlstone heteropic with turbidites of Marnoso-arenacea Formation, northern Apennines.

Plate 156
Horizontal burrows on a bed surface

These horizontal burrows are relatively common at the base of turbidite beds and are attributed to generic worm-like animals. They cross each other and current markings, suggesting that the organisms penetrated the sand bed from above and moved along the interface with the underlying mud. Similar structures could be produced by casting with sand the furrows left by reptating animals on a muddy bottom. A mold, however, would be a half relief, whereas the burrows seen in the picture are full-relief, cylindrical structures.

Marnoso-arenacea Formation, northern Apennines.

Photo: G. Piacentini 1970.

Plate 157
Helminthoid traces

The Linnean name of this structure (*Helminthoidea labirinthica*) reveals that it was regarded by paleontologists as a body fossil, although of uncertain classification (*incertae sedis*). The relief is modest, and the trace seems engraved on stone. Actually, it represents the track of an animal grazing a muddy bottom in a systematic way, more or less in the style of *Zoophycos* but without penetrating under the surface. Crossings and overlaps are absent, or almost so. The sediment is a pelagic marlstone, deposited at a bathyal or abyssal depth.

Helminthoids record the maximum depth of water (probably abyssal) among fossil traces, and are found in turbidite formations of the Alpine chains. These units, mostly of Late Mesozoic age, are collectively called "Helminthoid flysch" and crop out in tectonically displaced positions, in connection with compressive structures (thrust sheets). The compression produced by continental collision between Europe and Africa detached and uprooted piles of turbidites from their original substrate and formed the slices that are called thrusts (see also plate 1). A peculiarity of "Helminthoid flysch" units is that the turbidites are mainly calcareous, and individual beds often reach enormous sizes (megabeds). Intrabasinal carbonate detritus, formed by the microscopic tests of planktonic organisms, is supposed to have fed turbidity currents in the first place. The basement of the Helminthoid basins was probably constituted by oceanic crust (the western branch of the Tethys, which vanished at the end of the Mesozoic era or at the beginning of the Cenozoic).

Helminthoid Flysch, Cretaceous, western (French) Alps.

Plate 158
Problematic traces: *Chondrites*

Another case of "problematic" trace is illustrated here in a sample of thinly bedded marlstone. The size and the areal pattern are variable; the bladed shape, the local dendritic arrangement and the dark green color are reminiscent of vegetal remains. This explains why these pellicular traces were interpreted as fossil algae and given the name *fucoids*. They are now seen as horizontal burrows flattened by compaction. The color is related to a relatively high content of organic matter and silicates rich in Fe^{2+} (iron in reduced state), suggesting that the burrowing animals lived in a poorly oxygenated (disaerobic) environment. The burrows were probably excavated a few millimeters below the sedimentary interface.

The structure can be regarded as a *paleoceanographic indicator,* if it is assumed that the amount of oxygen dissolved in deep-seawater depends essentially on density-controlled oceanic circulation. Water density is regulated, in its turn, by temperature and salt content (salinity), whereby the movements of water masses in a vertical plane are qualified as thermohaline. *Chondrites* should reflect conditions intermediate between aerobic and anaerobic near, and immediately underneath, the bottom.

Fucoid Marls (Lower Cretaceous) of Umbria-Marche Apennines, Italy.

Photo: P. Ferrieri 1992. Scale is in centimeters. This and plates 159–162 are part of the collection of the Geology Department, University of Bologna.

Plate 159
Problematic traces: *Palaeodictyon*

Together with *Helminthoidea,* this honeycombed trace fossil is relatively common in turbiditic formations, although it was originally impressed not on turbidite beds themselves but on interbedded pelagic mud. This is demonstrated by the distortion of the hexagonal cells and the lacerations of the "net," which were caused by the arrival of a current (see also plate 93). The same current laid down the sand or silt the cast is made of.

All one can say about the animal responsible for *Palaeodictyon,* which is interpreted as a feeding trace, is that it had methodical habits, like *Zoophycos* and *Helminthoidea.* While at intermediate water depths, feeding strategies lead to shallow but still three-dimensional structures, deep-water traces seem to affect only a thin layer of sediment. Such superficial traces, or trails, plus some vertical burrows, are grouped in an assemblage of ichnofossils indicative of maximum depth in marine paleoenvironments, the so-called *Nereites ichnofacies.* Other assemblages characterize marine environments of intermediate depth (shelf to slope, with reference to Modern continental margins) and shallow depth (up to the intertidal and littoral zone), and continental environments, respectively.

In conclusion, trace fossils can be used as *paleobathymetric indicators,* but recognizing assemblages or associations is safer than relying on individual traces.

Photo: P. Ferrieri 1992.

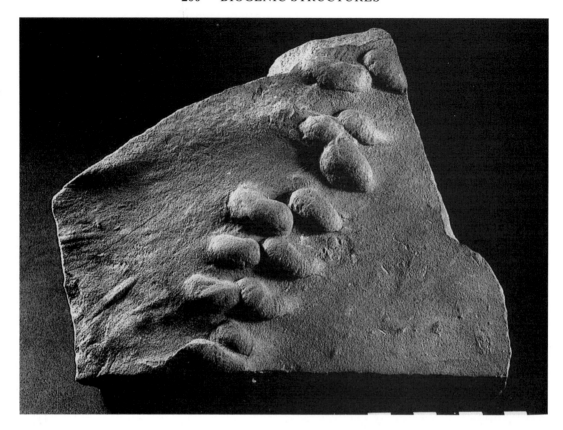

Plate 160
Problematic traces: *Neonereites*

This appreciable bas-relief, 3–4 cm wide, was more than a meter long in the original outcrop. As in previous examples, the trace fossil shows up as a cast on the sole of a turbidite sandstone bed, where it is associated with some current markings made by impacting "tools." Apart from these scattered structures, the original bottom was rather smooth.

The trace represents another member of the deep-water *Nereites* ichnofacies.

Marnoso-arenacea Formation, northern Apennines.

Photo: P. Ferrieri 1992.

(G. Piacentini, 1970.)

A

B

Plate 161
Traces of boring organisms

When organisms exert a boring activity, which is a chemical or mechanical attack, or a combination of them, scientists speak of *bio-erosion,* instead of bioturbation. Bioturbation is part of soft-sediment deformation processes, whereas bioerosion implies the disintegration or dissolution of hard or fragile substances (minerals, rock, wood). Dissolution affects mainly carbonate minerals, and is operated by acids secreted by organisms.

Plate 161 **A** shows perforated pebbles of various provenances, with holes of variable size. The pebbles were collected from beaches, where boring organisms (*lithophagous*) live mostly in the intertidal zone. They are represented chiefly by mollusks (e.g., *Lithodomus*), whose shells can still be found within the cavities. Bored pebbles are thus paleoenvironmental indicators, but attention must be paid to their potential reworking. Beach gravel can be involved in subaqueous slides or be captured by the head of a submarine canyon; it is then resedimented and becomes incorporated in deep-water deposits. The perforations are filled by sediment but remain preserved as witnesses to a former stay of the pebbles in the littoral zone (the gravel beach can be completely de-

stroyed, leaving only this indirect evidence of its existence). When streams empty their coarse load directly onto slopes and canyons, no bored pebbles are found.

Micro borings have been observed in thin sections of carbonate rocks; they are made by algae, fungi, etc. No documentation of these microstructures is presented here.

The strange structure of picture **B** results from the activity of wood-eating (*xilophagous*) organisms, in particular the mollusk *Teredo*. Their valves assume a tubular shape, which is preserved here by filling sand (fragmentary remnants of carbonate shell are still attached to the molds). The black speckles and strings indicate carbonized wood (lignite), and the surrounding rock is a turbidite sandstone (from the Marnoso-arenacea Formation). Apparently, the *Teredo* individuals were inside a piece of wood when it was carried in suspension by a turbidity current; very likely, they were still alive when sand deposition entombed them.

Minor structures, with the form and size of peas or pills, developed along the contact between the wood and the embedding sand. Their origin is uncertain (gas escape? liquefaction?).   *Photo:* **A** G. Piacentini 1970.

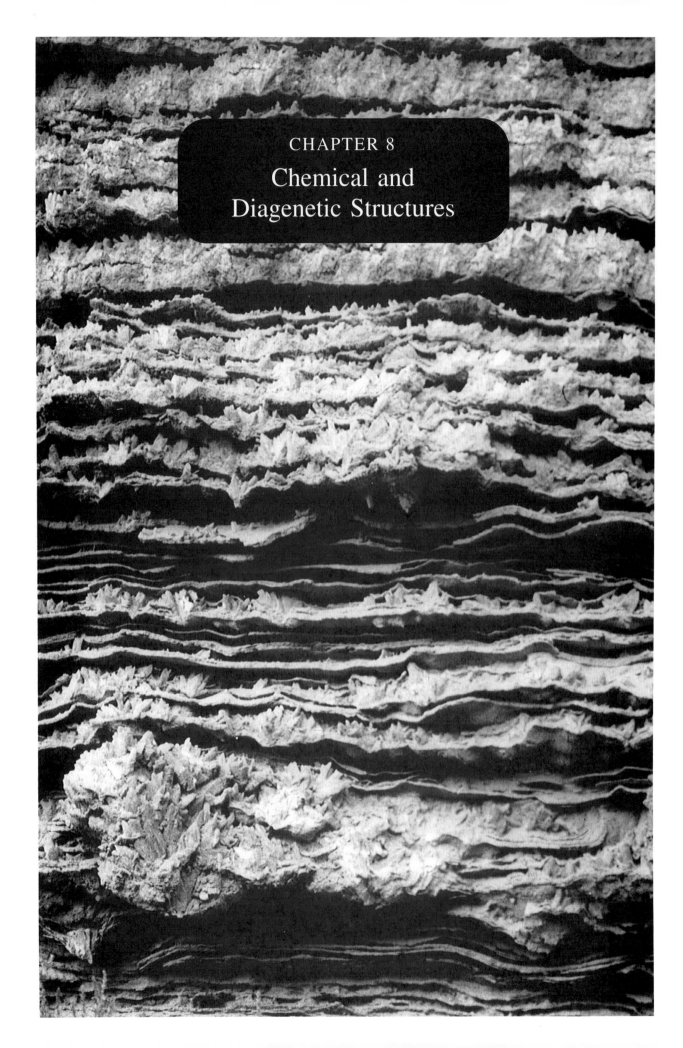

CHAPTER 8
Chemical and
Diagenetic Structures

The last group of structures we are going to examine includes *secondary structures,* whose origin is related to *diagenetic processes* acting beneath the depositional interface. Various mechanisms, both physical and chemical, operate during diagenesis; moreover, their effects vary in different materials. The review here is very synthetic and gives just a sample of significant categories. Within each category, the variety of phenomena is often very wide, depending on local factors and regional geological settings. A regional bias is, to a certain extent, unavoidable in presenting secondary structures.

It must also be remarked that the products of diagenesis are investigated chiefly in the field of textures, at a microscopic or ultramicroscopic (SEM) scale. At outcrop (mesoscopic) scale, they do not offer the same abundance of forms and geometries that primary structures display.

To be precise, not all structures of chemical origin are secondary; some of them are produced by or during deposition, and are included here, in the first part of the section (plates 162–166). Primary chemical precipitation, excluding evaporitic basins, is a local phenomenon; it occurs, for example, around springs, in lakes, rivers, etc. Similar products are found in karst environments, i.e., in large secondary voids below the topographic surface. Whether they are to be considered as depositional or secondary (by analogy with chemical cement lining or filling smaller pores) is largely a matter of preferences. Chemical structures in soils have, instead, the same aspect as diagenetic structures in recently buried sediments. In evaporitic strata, representing the bulk of chemical sediments (carbonates must be regarded mostly as *biochemical* products), both primary and secondary structures are common.

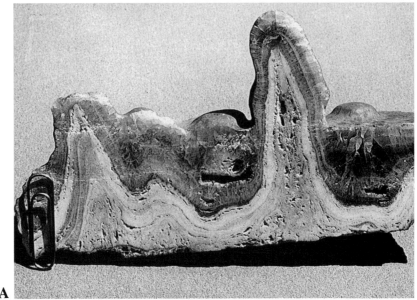

Plate 162
Primary chemical structures: limestone encrustations

Plate 162 **A** and **B** document two details of *speleothems,* as cave deposits are called. A stalactite has been cut, and the section shows two distinct generations of chemical precipitates, both consisting of calcium carbonate. The spongy inner zone (dripstone) is the result of dripping from the cave ceiling; the banded outer zone (flowstone) reflects, instead, the slow growth of crystals from solutions flowing *along* the solid surface. Flowstone deposition seals and drapes the dripstone, which implies a change in water circulation in the karst system. All cave "formations" (the term is used by speleologists for morphological varieties of precipitates, which geologists would call "structures": stalactites, stalagmites, basins, ponds, etc.) owe their origin to the release of carbon dioxide from water percolating through the karst system.

Carbonate ions are present in solution, and the escape of CO_2 force them to separate as a solid phase by decreasing their solubility.

In plate 162 **B** it can be more clearly seen that: 1) flowstone bands are sets of depositional laminae, recognizable by differences in color; 2) the laminae are made of elongated crystals, which are oriented normally to the lamina surface; 3) the parallelism and continuity of the laminae is remarkable; a small displacement is observable, however, along a fracture on the right (a micro fault). Flowstone laminae are possibly organized in seasonal rhythms (*varves*) and longer-term cycles, as indicated by minor and major color changes between individual laminae and laminasets, respectively. The color of pure carbonate is white; darker hues are due to variable

C

D

concentrations of impurities (iron oxides and hydroxides). The major changes in impurity content should reflect multiannual hydrologic-climatic changes.

Calcite crystals formed the laminae by growing side by side and disturbing each other; maximum growth then occurred perpendicularly in the free space of the cavity, which explains their elongated shape. This banded cave deposit is used as ornamental stone with the name of *calcareous alabaster*.

In plate 162 **C**, the carbonate encrusts vegetal remains. Precipitation can be restricted to the vicinities of springs or falls, or be more extensive and form continuos layers. *Travertine* deposits, or tufa, characterized by a spongy or vuggy texture, can thus accumulate. Their thickness can reach tens of meters; several travertine quarries are open in central Italy. The stone is used as a cheaper substitute for marble.

Plate 162 **D** shows an aggregate of calcareous *pisoliths*, spheroidal particles made of concentric shells of calcium carbonate. The internal structure is magnified in plate 163. Pisoliths grow in both external and subterranean ponds containing slightly turbulent water saturated in carbonate. Similar but smaller particles, called *ooliths* (or, better, *ooids*) are found in agitated waters of the marine realm, more precisely in marginal areas of carbonate platforms. They derive from stepwise coating of tiny suspended particles in CO_2- rich water currents.

All samples belong to the collection of the Department of Geological Sciences, University of Bologna.

Photos: G. Piacentini 1970.

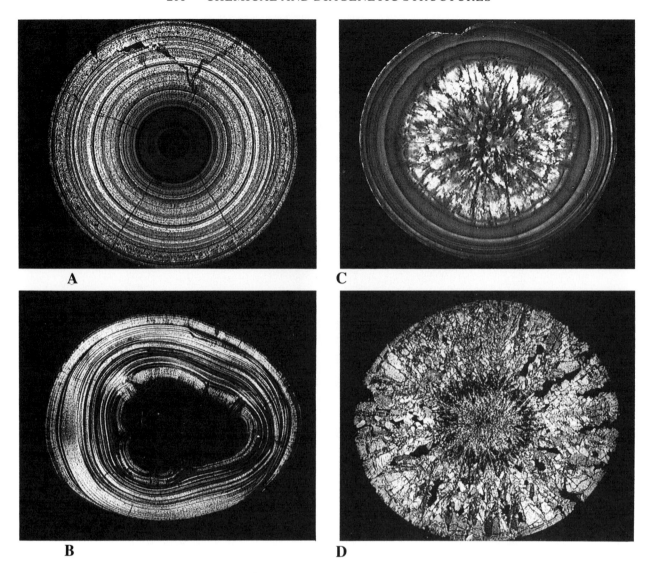

A

C

B

D

Plate 163
Sections of cave pisoliths

Plate 163 gives an idea of what pisoliths look like under the microscope, in transmitted light (thin section). Different microstructures and growth patterns can be appreciated. Two main fabrics stand out: 1) zoned-concentric, and 2) radial. Specimens **A** and **B** have a dark core of aragonite (less stable form of calcium carbonate), while the outer shell is made of calcite. In **B,** the core has a botryoidal shape and consists of aggregated particles.

The diameter varies between 2 cm (**A** and **B**) and 5–6 cm (**C** and **D**).

The concentric zonation of pisoliths indicates a discon-tinuous growth; longer pauses are manifested by micro-unconformities (see **B**), which imply that accretion was not continuous and the particles remained stationary for some periods. The constituent carbonate crystals are so tiny that cannot be resolved. In the radial pattern, on the contrary, they are much larger. This means that a secondary reorganization (a *recrystallization*) has occurred, which led to an enlargement of the original crystals.

The pisoliths were collected in caves within the karstified Messinian gypsum near Bologna, Italy.   Photos: G. G. Zuffa 1970.

Plate 164
Accretionary lapilli

Partly resembling pisoliths, these spheroidal particles are found in volcanic ash deposits and owe their origin to completely different mechanisms. Properly speaking, they do not represent chemical structures, but are presented here for comparison purposes and also because pyroclastic deposits are emphasized in this book.

As opposed to ordinary lapilli, which are clasts (see white pumice fragments in plate 164 **B**), accretionary lapilli are friable aggregations of fine ash particles, once attributed to raindrops. The most plausible explanation for their formation lies in the condensation of water vapor within a highly concentrated, hot pyroclastic flow or an ascending eruptive column. Water droplets acted as aggregation nuclei by making the ash particles stick together. Vapor is abundant in phreatic eruptions, occurring when the volcanic heat vaporizes a water body (aquifer, lake). There is no cement in accretionary lapilli, which can be crumbled between two fingers.

The examples shown here crop out in Quaternary deposits of Eolian Islands, Tyrrhenian Sea.

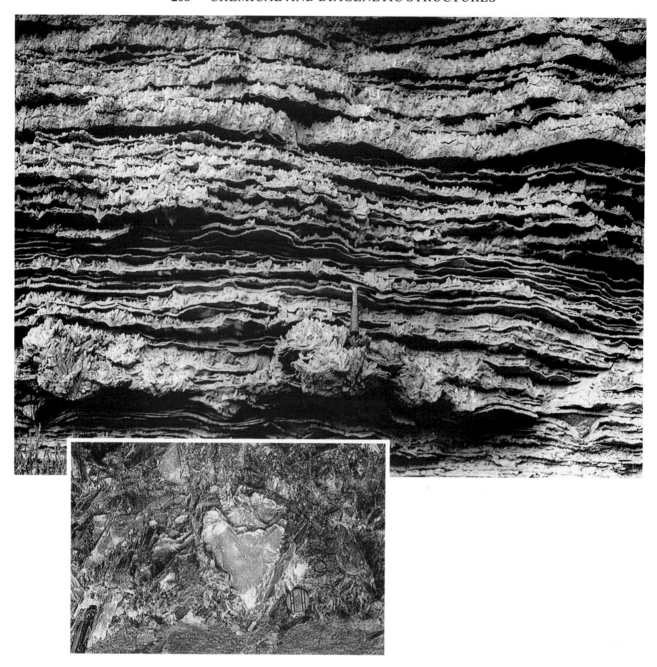

Plate 165
Crystal aggregations: palisades and bushes (*cavoli*)

Gypsum beds protrude from the outcrop and alternate with pelitic partings. Part of the gypsum is finely crystalline, thin-bedded and laminated (*"balatino"* facies, as it is called in Sicily), part is macrocrystalline (selenite) and organized in laterally continuous beds (upper part of the section), incomplete beds (center), and isolated clusters, or "bushes." The bushes, or *cavoli*, deformed underlying beds under their weight. The continuous layers are called *palisades* for their flat base (the substrate where crystal growth started) and ragged top, which reflects the tails of the selenite twins (see detail in the small photo inset, and

plate 142). Crystal growth is influenced by the number of seeds, or nuclei, per unit area. Where the nuclei are closely spaced and uniformly distributed, lateral growth is restricted and a vertical alignment ensues. In clusters, crystal axes diverge upwards because lateral confinement is reduced or absent.

Selenite crystals nucleate within shallowly submerged to intertidal sediment, and can incorporate bacterial mats during their growth (see plate 142).

Messinian gypsum beds of Eraclea Minoa, Sicily.

saturated solution   diluted solution

fondo fangoso

1 Precipitation   2 Crystal growth   3 Dissolution   4 Clastic fallout

Plate 166
Salt crystal casts

The samples were taken from the base of a siltstone bed in a Triassic dolomitic formation (Muschelkalk of central Europe). The marks on the bed surface reveal the shape of cubical objects, which have been replaced by silt sediment. These objects were crystals of halite (NaCl), precipitated from an evaporitic brine (hyperconcentrated solution) on the soft muddy bottom of a lagoon or a natural salina. Contrary to gypsum crystals, which nucleate within a host sediment, halite crystals form at the surface of the water body and fall to the bottom after reaching a certain size (their weight must win the surface tension of the brine). The crystals keep on growing on the bottom if the solution is supersaturated there. If a dilution occurs, the salt is dissolved but leaves its marks on the mud. Such a dilution may be caused by an influx of fresh water from rain or streams, or of marine water of normal salinity from the ocean.

Halite is one of the more soluble salts present in seawater and several lakes, and is extremely sensitive to salinity changes. At least 90% of water must evaporate to make halite separate from the solution.

One can hypothesize that the water of the Ancient lagoon evaporated completely and the bottom dried up. In that case, even a small amount of fresh water could have dissolved the salt. Is this hypothesis compatible with what you observe in the picture?   *Photo: G. Piacentini 1970.*

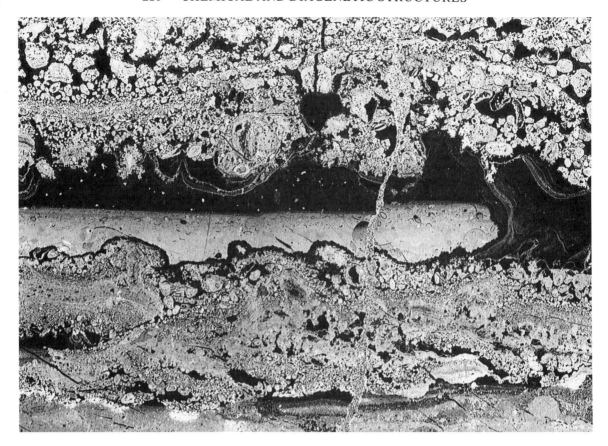

Plate 167
Geopetal structure

This curious name was given to structures that help to determine the stratigraphic polarity of beds (way-up criterion). In some cases, they are visible to the naked eye, but are more commonly microscopic. The picture shows exactly a microscopic example in a thin section of a carbonate rock (black and white parts are reversed because of the negative print).

A geopetal structure consists in the double filling of a cavity inside the sediment. The first phase of filling is represented by deposition of fine clastic particles that are filtered by the coarser framework of the sediment. This *internal sediment* is horizontally layered under the control of gravity. Later on, the remaining void is occupied by chemical cement, whose crystals seal the internal sediment; its upper surface then becomes a fossil level, i.e., an indicator of *paleohorizontality*. Geopetal structures are very useful in poorly bedded rocks like carbonate reefs and banks, where internal cavities are frequent.

The rock of this example is a calcarenite, i.e., a lithified carbonate sand from the Calcari Grigi Formation of southern Alps.[1] The cavity is a *leaching vug,* caused by dissolution of the already lithified carbonate (see irregular, corroded contour). It is, in practice, a microkarst feature, a kind of miniaturized cave.

Plate 168
Stromatactis structure

This structure is found in carbonate rocks, mostly of Paleozoic age, but a Recent analog has been reported from the Florida Channel. Its origin is not yet fully understood and has long been debated.

Stromatactis is the name given to irregularly shaped masses of sparry calcite (or sparite, made of clean, relatively large crystals) scattered in, or alternating with microcrystalline calcite (micrite). In the picture, micrite forms the darker mass. The contours of the sparite spots are sharp and jagged.

According to one of the most credited interpretations, the calcite crystals filled leaching vugs that originated in moundlike bodies of semilithified carbonate. More precisely, the mounds would have built up in shallow water by a steplike process, with alternating layers of loose sediment (carbonate mud) and crusts of early lithified mud. The fine carbonate particles were probably aggregated by some kind of microscopic organisms (algae, bacteria, or others). Later dissolution would have selectively removed the mud horizons. The crusts would have locally collapsed for lack of support, and this resulted in a series of irregularly distributed cavities. The last phase (the "seal") was represented by the filling of the cavities by chemical cement.

A Lower Devonian stromatactis-bearing carbonate mound from southern Sardinia.

Photo: M. Gnoli et al. 1981. Neues Jahrbuch Geol. Paleont. Monatshefte H-6.

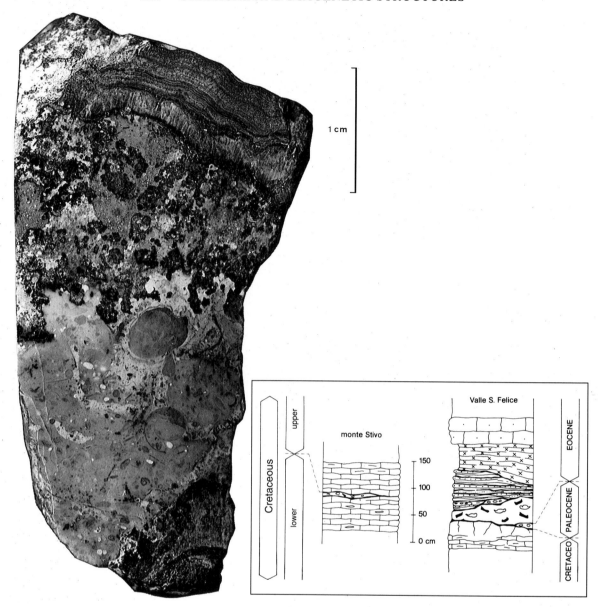

Plate 169
Hardground structure

A hardground is a submarine crust of indurated sediment, in practice, a horizon that has been lithified before burial, or pre-consolidated. Its structure is a compound of *duri-crust* and nodular structures. A hardground is, therefore, a complex object, and various processes contribute to its formation: dissolution and reprecipitation of calcium carbonate, precipitation of iron and manganese oxides, biological activity of boring organisms, etc. They can operate if a basic requirement is met: very slow or no sedimentation. This condition is provided in pelagic environments (open sea, far from land influences), when organic productivity is drastically reduced (the hard parts of dead plankton raining from the surface waters represent the main contribution to sedimentation) or currents rich in CO_2 sweep and corrode the bottom.

Given these circumstances, hardgrounds are indicators

of stratigraphic gaps (hiatuses) and discontinuities. Examples are familiar in Mesozoic and Early Tertiary carbonate formations of the Tethys (Mediterranean-Alpine) area. Plate 169 shows one of them, in the Upper Cretaceous Scaglia Formation of southern Alps.[2]

A magnified polished section is here placed in upright stratigraphic position. An outer crust, made of several bands or laminae as in speleothems and continental duri-crusts (see plate 162), caps a mottled zone where remnants of the original sediment (lighter patches) intermingle with nodules of phosphate and metal-rich precipitates. The unaltered sediment, a micritic, porcellanaceous limestone, appears in the lower part of the section. It contains abundant fossil shells, concentrated by selective dissolution of sediment.

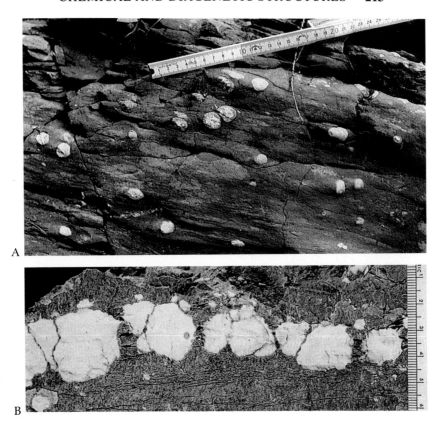

Plate 170
Nodular structure: isolated nodules

The white specks in these images are nodules, i.e., aggregates of small crystals precipitated within a sediment permeated by fluids. The interstitial fluids do no consist of water only but also of air and/or biogenic gases; neither is the water pure, but contains various ions, or salts, in solution. In plate 170 **A,** for example, the nodules are made of calcium carbonate, in **B** of calcium sulfate (anhydrite). In both cases, the host sediment is continental (alluvial) and has a red color. In humid climates, the topmost part of these deposits, exposed to the atmosphere, is *pedogenized,* or transformed into soil, with the cooperation of water and various organisms. In drier conditions, pedogenic processes cannot proceed in the usual way; only some physical and chemical modifications can occur.

Among them, the dissolution and reprecipitation of soluble substances when water is available (which happens for short periods). In this respect, the formation of nodules and crusts represents rather a form of diagenesis (*vadose diagenesis*) than of pedogenesis. The strong evaporation of arid and semiarid climates "pumps up" water that infiltrated underground during rainstorms; the water is forced upward by capillarity, and salts separate from it. The precipitation of carbonate is favored not only by evaporation but also by the alkalinity of the solution (acid substances are lacking). The carbonate crusts that form just below the topographic surface are called *caliche, calcrete, kankar,* etc., in different regions of the world. They must not be confused with ordinary beds, i.e., primary deposits. The red color of the sediment is related to oxidation of iron.

Sulfate nodules and crusts indicate extreme aridity, as the mineral is evaporitic; they form in environments known as *sabkhas*. Inland sabkhas are located in the sub-bottom of ephemeral (*playa*) lakes, coastal sabkhas in barren, salt-encrusted flats exposed to the sea.

Both the carbonate and the sulfate in the nodules shown here are almost pure, as indicated by their white color. This means that the growth of nodules was "displacive": the particles of the host sediment were not accepted inside the structure, but pushed aside. The nodules had, originally, a porous texture, which was subsequently occluded by the hydration of anhydrite into gypsum. Morphology and size of the nodules did not change when gypsum replaced anhydrite.

In conclusion, nodular structures related to vadose diagenesis are *paleoclimatic* indicators pointing to emergent and semiarid or arid conditions. They should not be confused with nodules of other origins (compare with plates 169 and 176). *Photos:* **A** *G. G. Ori 1992;* **B** *A. Bosellini 1992.*

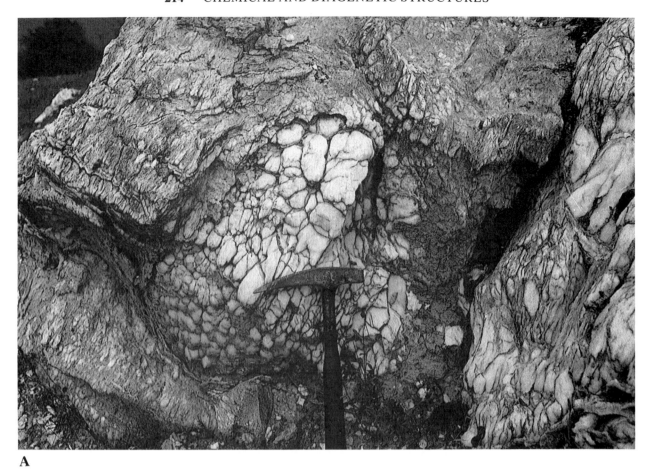

A

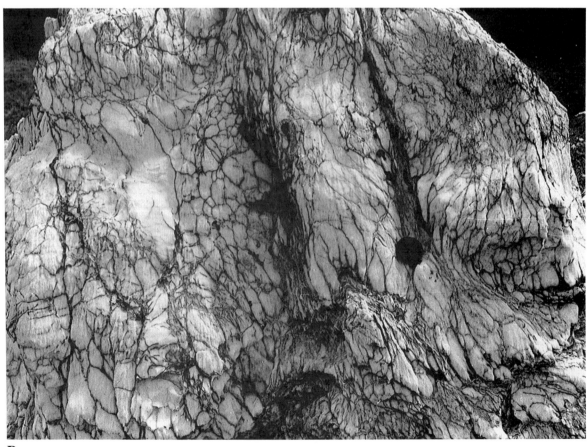

B

Plate 171
Nodular structure: composite nodules (A) and chicken-wire structure (B)

Single anhydrite nodules resulting from vadose diagenesis can aggregate into composite nodular masses of roughly equal dimensions (**A**) or coalesce laterally into "diagenetic beds" or crusts (**B**), which can be deformed by enterolithic folding (see plate 127). Crystalline, saccharoidal gypsum that replaces anhydrite when conditions become wetter is called *alabastrino,* from the old name "alabastro" given to gypsum quarried near Volterra in Tuscany, Italy, where the outcrops pictured here are found. Gypsum alabaster must not be confused with calcareous alabaster (see plate 162 **A, B**).

The nodules seen in plate 171 **A** are embedded in a groundmass of clastic gypsum particles mixed with terrigenous sand and clay. This means that they have been reworked by mechanical agents (gravity, storms, etc.) and transferred to deeper parts of the basin. No surprise, then, if one finds indicators of subaerial conditions in subaqueous sediments. Remember that only biogenic structures (trace fossils) cannot be found in reworked (allochtonous) position, simply because they are destroyed upon removal by erosion, slides, etc.

In composite nodules, the host sediment is reduced to thin films and the individual nodules are tightly packed, which is called chicken wire, or mosaic structure. This structure can invade considerable masses of sediments, where it destroys any trace of bedding (**B**).

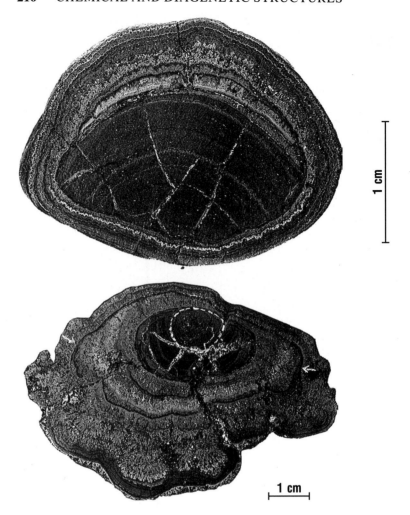

Plate 172
Sections of manganese nodules

Nodular structures do not occur only in continental sediments and soils. They are found in marine deposits, too: *chert nodules,* for example, are common in pelagic limestones of European Mesozoic successions, both in well-lithified formations of the Mediterranean region and the famous Chalk of the Paris Basin and the English Channel (the "white cliffs of Dover").

Manganese nodules pave vast expanses of the deep ocean floor; the samples pictured here were dragged from a depth of 5,000 m in the Pacific.[3] They rest on very fine clays of red or brown color, whose rate of accumulation is extremely low; the nodules grow even more slowly (less than 1 mm per thousand years) but nonetheless are only half-buried by the clay. The are probably moved from time to time by currents or organisms searching for food.

Manganese is probably extracted from seawater by microorganisms and fixed around a tiny nucleus. The coating is not symmetrical, and the internal structure of

nodules is eccentric. This means that the growth is not a continuous process, as confirmed by the discordant contact between sets of laminae. Moreover, it is slower on the upper side, as can be seen from the greater thickness of the lower, buried side. This suggests that the manganese supply comes mostly through interstitial water. Together with manganese, other metal ions (cobalt, iron, platinum, nickel, etc.) are precipitated in the nodules, which should be called *polymetallic*. The concentrations are high and would make the nodules an economic resource were it not for the formidable technical and environmental problems that their mining poses.

Red abyssal clays are a typical *residual sediment,* concentrating insoluble impurities of calcareous and siliceous organic particles that are dissolved below a certain depth. Carbonate skeletons, in particular, are attacked by slightly acidic, CO_2-rich abyssal waters, and do not "survive" below a critical limit, the CCD (carbonate compensation depth).

Plate 173
Arenaceous concretions

Globular masses often stand out in outcrops of poorly cemented sandstones; they are secondary features, whose occurrence does not depend on the depositional setting. Here, for example, similar forms are shown in littoral (**A**) and turbiditic (**B**) deposits. These *concretions* are different from the nodules examined in previous plates under a fundamental respect: they are not displacive, but inclusive. In other terms, the host sediment is not pushed apart by the growth of the concretion, but is included in it. Carbonate cement is simply precipitated in the pore spaces.

Whether precipitation was diffused and simultaneous in the concretion volume or expanded radially from a nucleus, it is not clear. If the concretion is split apart, sometimes an object is found in the core (a clay or a wood fragment, a shell or a lens of skeletal debris), sometimes

nothing appears. Why, then, cementation is so spotty remains a mystery. Much more so because other examples of selective cementation, more easily explainable, occur in the same sandstone units, for example along bedding or lamination surfaces (see plate 41). Stratiform cementation is particularly developed near permeability barriers, like mudstone partings. It cannot be considered, however, as a real structure.

Whatever their origin, arenaceous concretions are useful for the field sedimentologist, as they emphasize primary structures that would remain scarcely visible in the surrounding sand (see **B**).

A: Pliocene nearshore deposits of the Intra-apenninic Basin; **B:** Upper Marnoso-arenacea Formation, Santerno Valley, northern Apennines.

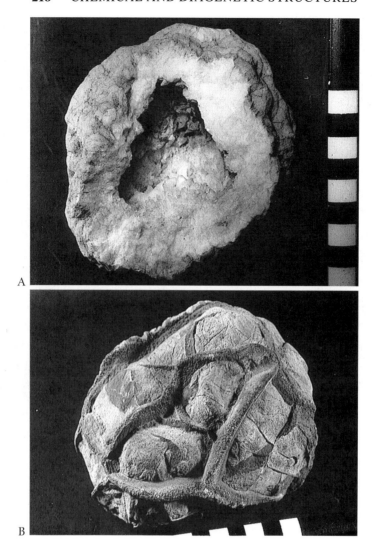

Plate 174
Hollow concretions: geode (A) and septarian "nodule" (B)

A *geode* is an isolated cavity lined by crystals that grow inward; a *druse* is an open, sheetlike cavity, or fissure, also coated by crystals. The crystals are usually macroscopic, with regular faces developed on their free ends: in the example shown here, coming from a sulfur mine in Messinian evaporites, the mineral is celestine (strontium sulfate). When the crystals are very small, the coating is in the form of bands or crusts of different thickness and colors draping the cavity walls.

The concretion shown in plate 174 **B** is called *septarium* (plural: *septaria*) because of its external aspect. The interior, not shown, consists of a cracked concentric layer (cavity coating, with inward widening cracks), sometimes with a breccia at the core (due to collapse of fragments from the coating). The cracks and the residual cavity are partially or totally filled by a second generation of chemical precipitates, usually in the form of macroscopic crystals. The second generation mineral is harder than the first one, which is removed more easily by weathering; that is why the crack fills appear as ridges (*septa*) on the external surface, with a typical reticulate pattern.

All these concretions are, originally, gel-like masses in which dehydration and crystallization occur with time; the loss of water causes shrinkage and the opening of a cavity. Before hardening, a hollow concretion cannot withstand an overload, which implies that crystallization occurs under a shallow burial depth.

The samples are part of the collection of the Department of Geological Sciences, University of Bologna. Scale in centimeters.   Photo: P. Ferrieri 1992.

Plate 175
Rosette-shaped gypsum concretion

This large and spectacular concretion (see watch on left for scale) has a discoidal shape and a radial growth pattern, the "rays" being made of saberlike selenite crystals. The crystals are turbid for the presence of clay impurities. Similar concretion, called *rosettes* or *cockades,* occur more commonly on a much smaller scale (millimeters to centimeters). They are found along joints and bedding planes (i.e., physical discontinuities in a rock mass), which explains their flat shape.

This gigantic example was collected in Pliocene fossiliferous clays of the Siena area in central Italy. The obvious conclusion is that this diagenetic form of gypsum has nothing to do with evaporites, unless sulfate ions deriving from a leached evaporitic formation circulate through the rocks. But it is not necessary to have a sulfate source: the SO_4^- ions can derive from the oxidation, possibly helped by bacteria, of sulfides (which are commonly contained in normal marine muds). The calcium is supplied by the carbonate skeletons of fossils.

Joint-filling gypsum is usually made of elongate, fibrous crystals oriented perpendicularly or obliquely to the joint surfaces (see plate 176 and color picture 29); this variety of secondary gypsum has a white color and a silky gloss, and is called *sericolite*. One can wonder why rosette concretions, too, are not made of sericolite. Actually, they were. Only afterward, sericolite recrystallized into twinned selenite; fibrous crystals, forming a double, welded palisade, are still recognizable near the center of the concretion. *Photo: A. Ferrieri 1992.*

Plate 176
Gypsum veins

The final two plates illustrate examples of structures that can be regarded as tectonic rather than sedimentary. One can thus have a taste of differences and analogies, especially in the fields of deformations (soft-sediment versus hard rock) and diagenetic processes (early vs. late diagenesis).

Veins are fractures and joints welded by crystals precipitated from circulating solutions. The mineral can vary depending on local factors: carbonates and quartz are the most common, but in this case there is gypsum, which is connected with sulfur-rich rocks.

Fracture and joints are produced in brittle rocks and sediments, either by the direct application of tectonic stresses or by a *relaxation of masses* that are relieved from previous stresses and loads. In open fractures, normally caused by extension, mineral solutions can circulate freely, whereas pressure is needed to force fluids along surfaces of shear.

In the outcrop of laminated gypsum shown here, veins are marked by white lines and form two systems: one is parallel to the depositional laminae, the other cuts across them. Transversal veins represent the fill of open joints, whilst the others result from injection of sulfate solutions under pressure; this "hydraulic fracturing" chose the lines of lesser resistance in the rock. Can you, applying the principle of intersection, determine which system is older and which is younger?

Other gypsum (sericolite) veins can be seen in color photo 29: they cross a shale bed separating two beds of evaporitic gypsum, and are folded and displaced along a tectonic shear plane.

Veins cannot be easily mistaken for sedimentary features: they sharply truncate primary and diagenetic structures as well. There is another secondary structure that truncates previous ones: the *stylolite*, or *stylolitic surface* (see color photo 31). In rock sections, it appears as a jagged line, resembling a cranial suture and marked by red or brown coats of iron oxides (no crystals, then). It can be seen, in our example, that stylolites are antecedent to some vein systems, posterior to others. This suggests that stylolites can be either pretectonic or tectonic features, which seems confirmed by the fact that some of them are about parallel to bedding (i.e., perpendicular to the load exerted by the sediment pile), and some have various orientations that reflect tectonic stresses. Actually, the majority of stylolitic surfaces are related to loading effects, and can be regarded as full title sedimentary structures (of deformational type).

Stylolites are produced by dissolution of soluble materials (typically, carbonates) under a solid pressure (a process named *pressure-solution*); their indented geometry indicates that the pressure is more focused on certain points, less on others. The oxides represent insoluble residues that remained in place (while the solutes were removed). A certain critical load and depth of burial are needed for stylolites to form; escape routes for solutions, through pores, fractures or the stylolitic surface itself, are also required.

Plate 177
Vein systems

Plate 177 shows the surface of a "cellular" or reticulated limestone, where the cells are outlined by crossing systems of veins. Veins are secondary structures due to precipitation of salts within fractures. This chemical cement is often harder than the fractured rock, which is removed first by weathering and erosion enhanced here by salt and wind action on a seashore. An inversion of relief thus occurs: cavities become protruding parts, as in the already discussed case of septaria. The trellislike patterns eventually end up in open frameworks through the complete removal of the original rock.

Paleozoic limestone, New South Wales, Australia.

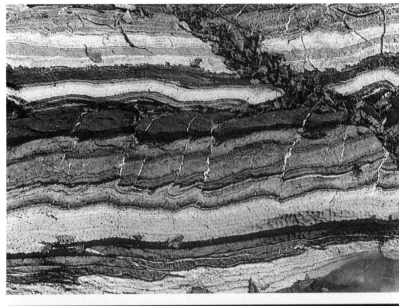

A

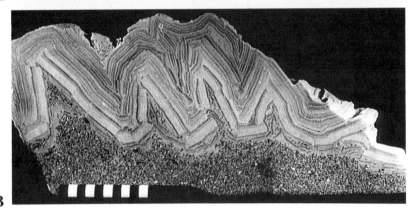

B

Plate 178
Tectonic deformations in sulfate evaporites

Evaporitic sediments are often indurated and brittle like crusts since their origin; even limited phenomena of dissolution and reprecipitation can weld their crystalline framework before they are buried. In other words, they constitute *preconsolidated* materials. When, upon burial, they are loaded, their behavior becomes ductile, i.e., similar to that of a plastic or semifluid substance. Owing, in fact, to the solubility of salts and the presence of hydrated minerals, a lot of water can be squeezed from evaporitic bodies and lubricate their movement. Two causes of deformation can be identified: gravity (load), and oriented tectonic stresses. The first cause induces *diapirism,* consisting in upward extrusion of salt masses (domes); more precisely, unequal loading of buried salt strata and density inversion with respect to overlying sediments (whose consolidation increases with depth of burial while evaporites, being preconsolidated, do not change in density) are the main causal factors of this form of instability.

Salt domes are large-scale features; at smaller scales, the deformation of evaporites is represented by intricate folds and flowage structures, partly similar to convolutions of clastic sediments. A folding style is also the response to tectonic stresses, as documented by the two pictures in plate 178, which illustrate compressed rocks

(**A** from Messinian gypsum and oil shales of the Apennines; **B** from Triassic anhydrite and dolomite of Germany). The vergence of the small folds, in both cases, indicates the direction of relative movement that resulted from a shear couple, or rotational shear (the upper parts are displaced more than the lower parts).

Noteworthy is some evidence of brittle behavior in both examples: in **A,** some beds are crossed by veins, in **B** dolomite laminae interbedded with anhydrite are fractured and "kink-folded" (hinges of fold). The sericolite veins in **A** show small-scale (*ptygmatic*) folding, which suggests fluid injection rather than fracture-filling and calls into question the rigidity of the mass.

Photo: P. Ferrieri 1992.

ENDNOTES

1. A. Castellarin and R. Sartori. 1973. Desiccation shrinkage and leaching vugs in the Calcari Grigi Infraliassic tidal flat. *Eclogae Geol. Helvetiae* 66: 339–343.

2. A. Castellarin, F. Frascari, and M. Del Monte. 1974. Cosmic fallout in the "hard grounds" of the Venetian Region. *Giornale di Geologia* 39: 333–346.

3. U. Von Stackelberg and V. Marchig. 1987. Manganese nodules from the Equatorial North Pacific Ocean. *Geol. Jahrbuch* D87: 123–227.

Colors, Forms, and Lines in Sediments

1. Geometry of sedimentary bodies: carbonate buildup (fossil reef). Permian Trogkopfel Limestone, Carnic Alps.   *G. B. Vai, Bologna.*

2. Geometry of sedimentary bodies: foreset bedding in Ancient fan delta slope. Evrostini, Greece.   *G. G. Ori, Bologna.*

3. Wavy lines in bedding: ash layers deposited by pyroclastic flow and fall. Island of Procida, Tyrrhenian Sea.

4. Wavy lines in bedding: tractive laminae and bed forms in surge-flow pyroclastic deposits. Island of Procida, Tyrrhenian Sea.

5. Small-scale undulations: hummocky cross-lamination semi-obliterated by burrowing organisms in Pliocene, transgressive calcareous sands. Scoglio del Trave near Ancona, Adriatic coast.

6. Small-scale undulations: laminated overbank sediment disturbed by water escape and plant rootlets. Paleozoic redbeds, New South Wales, Australia.

7. Parallel, wavy and crossing lines: amalgamation of two turbidite sandstone beds, with load lobes and flames along the contact (upper right). In the lower bed, three levels (a, b, and c divisions of the "Bouma Sequence") can be recognized, from base (left) to top: structureless sand, plane-parallel lamination (with dark, carbonaceous laminae), and cross-lamination with climbing foresets. Upper Miocene Laga Formation, near Ascoli Piceno, central Apennines.

M. A. Bassetti, Bologna.

8. Wavy forms: capitol-shaped pillow structure (pseudonodule) showing "toothpaste" style of deformation. Silurian Thorold Formation, New Jolly Road Cut, Hamilton, Ontario.

13. Surface forms: a star-shaped dune, Sahara desert.   *R. Sartori, Bologna.*

14. Crossing lines: giant cross-bedding in Jurassic Navajo Sandstone, Zion National Park, U.S. Fossil equivalent of eolian dunes.

15. Crossing lines: cross-bedding in pyroclastic deposits: migrating dunelike bed forms in a Quaternary tuff ring, Australia.

16. Crossing lines: climbing ripple cross-lamination in fluvial sands (plastic replica). Scale in inches.

Courtesy of J. B. Southard.

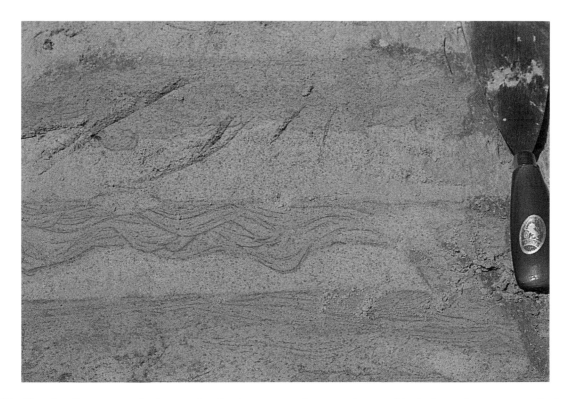

17. Crossing lines: wave-ripple cross-lamination on top of structureless sand beds (normal waves reworked storm deposits). Pleistocene deposits of Apennines foothills, near Ancona, Italy.

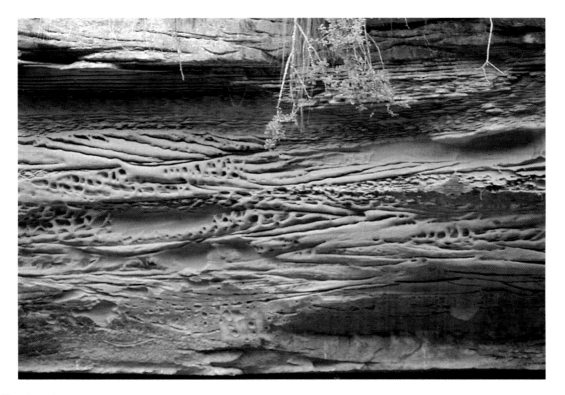

18. Crossing lines: thin mud drapes on ripple and dune scale cross laminae. Oligo-Miocene Swiss Molasse near Fribourg, Switzerland.

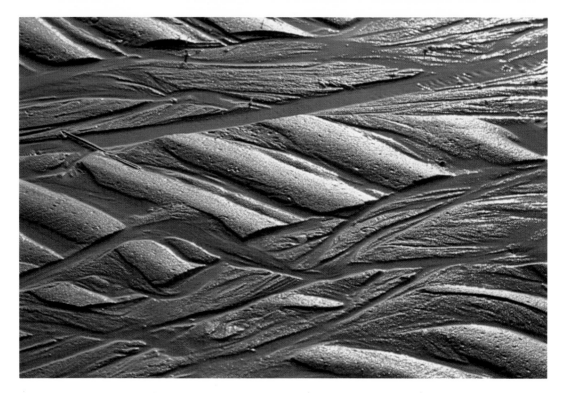

19. Crossing lines: wave ripples re incised by rill marks on a beach. Adriatic coast near Ravenna, Italy.
V. Rossi, Bologna.

20. Crossing lines: injection of liquefied sand deforming tractive laminae. Miocene fluvial deposits, Spain.

21. Circular lines: plan view of a stromatolite mound. Paleozoic carbonates of Petrified Sea Gardens, N.J.
A. Ferrari, Bologna.

22. Circular lines: swing mark in eolian sand.   *V. Rossi, Bologna.*

23. Circular forms: erosional moat and diverging eolian ripples around a fixed obstacle.   *V. Rossi, Bologna.*

24. Circular forms: spiraled trace fossil (*Zoophycos*) in Carboniferous Limestone of Scotland.

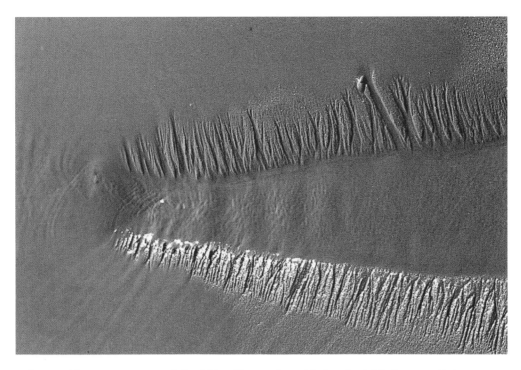

25. Linear forms: ebb current scour and dendritic rills on a low-tide beach. Adriatic coast, Italy. *V. Rossi, Bologna.*

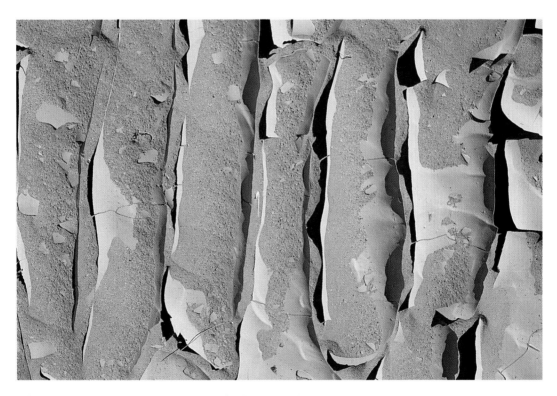

26. Linear forms: symmetrical wave ripples covered by a desiccated mud film and windborne sand. Note incipient mud curls.

27. Chaotic forms: broken and curled polygons in desiccated mud.

28. Forms related to crystal growth: "chaotic" aggregates of ice crystals on a mud surface.

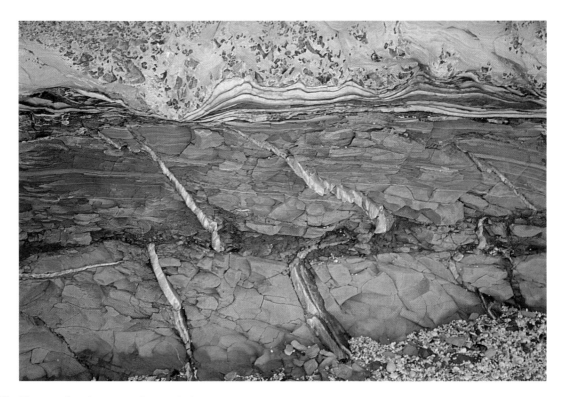

29. Forms related to crystal growth: load casting at the base of a gypsum bed, and deformed veins of fibrous, secondary gypsum (sericolite) crossing bituminous shale. Messinian Gessoso-solfifera Formation, Il Trave near Ancona.

30. Forms related to crystal growth: tufts of selenite crystals with organic impurities in a groundmass of clastic gypsum (gypsarenite). Messinian Gessoso-solfifera Formation, Il Trave near Ancona.

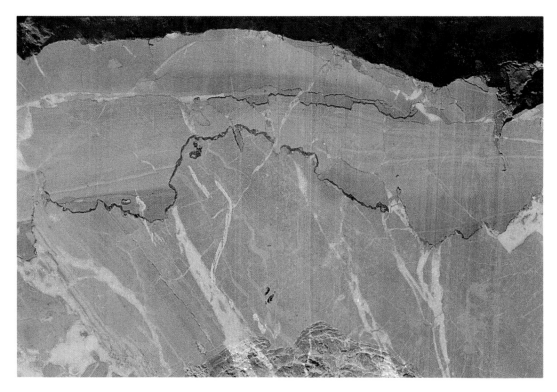

31. Forms related to lithification: red-stained stylolites cutting calcite veins in a Jurassic pelagic limestone (Ammonitico Rosso), Dinarides, Balkan Peninsula.

32. Forms related to diagenetic processes: nodules of amorphous sulfur in a gypsum bed. The sulfur derives from bacterial reduction of the calcium sulfate.

Glossary

Note: See index for references not found here.

accommodation space: space made available for sedimentation below water level.

adhesion ripples: small-scale structures produced by wind blowing sand on a wet surface.

aeolianite: *see* eolianite.

agglomerate: conglomerate composed of mixed volcanic derived materials.

aggradation: vertical accretion of sedimentary interface; *cf.* lateral growth, progradation.

air-heave structure: distortion of depositional structures (e.g., laminae) by ascending gas.

aklè dunes: fields of barchan dunes; *see* color photo 11.

algal biscuits: old term for algal (or bacterial) encrusted grains; *see* oncoid, oncolith, and rhodolith; *also see* plate 144.

algal lamination (structure): synonymous with the more modern term *bacterial* lamination (structure); *see also* algal mat, oncolite, rhodolith, stromatolite.

algal mat: should be updated into *bacterial* or microbal mat.

allogenic: attribute of material deriving from outside the basin or the region; *syn.:* extraformational, extrabasinal, exotic, allochtonous, allotigenous, terrigenous.

alluvial: pertinent to fluvial deposits.

amalgamation surface: subdued bed surface of erosional origin separating two coarse-grained beds; erosion caused the welding of two similar or identical lithotypes.

anastomosing ridges: small-scale markings made by currents on the surfaces of sand or silt laminae (intrastratal or parting-plane structure).

anastomosing rill marks: morphological variety of rill marks: *see* color photo 19.

angle of repose: critical angle of slope of a loose sediment, representing an equilibrium condition between gravity (tangential component of weight) and internal resistance to shear (angle of internal friction).

antidunes: tractive bed forms formed in supercritical flow regimes; *see* plate 37.

antidune phase: range of hydraulic conditions favorable to formation of antidunes.

antiripples: *see* adhesion ripples.

appositional fabric: spatial arrangement of sedimentary particles following deposition; *cf.* imbrication.

arcuate bands: intersections of foreset laminae (cross-lamination) with the base of the bed; *syn.:* rib-and-furrow structure.

armored mudballs: sand or pebble coated mud clasts; *see* plate 80.

asymmetrical ripple marks: small-scale bed forms produced by a unidirectional current or waves with a dominant component of oscillation.

asymmetry index: descriptor of ripple asymmetry, equal to the ratio between the lengths of the two sides.

attachment point: where flow lines come in touch with a solid boundary after losing contact with it before or behind an obstacle; down current limit of a separation bubble; *cf.* separation point.

auto-breccia: breccia formed during lava consolidation, by rapid quenching or detachment of consolidated portions and their incorporation in the still fluid mass.

avalanching: sliding of groups of grains on the slipface of a bed form.

backfilling: retrogradational filling of a channel or scour; *see* plate 75 **A.**

backreef: internal, protected part of an organic reef, usually occupied by a lagoon.

backset beds (laminae): up current dipping beds (laminae).

backshore: emerged part of a beach; *cf.* foreshore; shoreface.

ballistic ripples: eolian ripples produced by saltating sand grains.

barchanoid (ripples, dunes): *see under* lunate.

base-absent sequence: bouma sequence lacking one or more lower terms; *cf.* bouma sequence, turbidite,

base surge: type of pyroclastic flow (turbulent suspension), driven by gravity and volcanic blast.

beach cusps: series of small "promontories" alternating with rounded embayments, which form along a beach (foreshore) in the wake of storms.

bedding plane structures: *see* external structures, interfacial structures.

bedform lag: delay between the discharge peak of a tractive current and the maximum development of related bed forms.

bedload: the bulk of particles transported near the bottom by a current.

bedset: packet or bundle of conformable beds.

berm: shallow ridge, bench, or terrace, separating foreshore from backshore.

bioconstruction: building up caused by active growth of colonial, sedentary organisms.

biodeformation: *see* bioturbation.

bioglyphs: *see* ichnofossils, trace fossils.

bioturbation: mechanical disturbance and deformation of sediment induced by organic activity.

bird's-eye (birdseye) structure: internal cavity related to desiccation, leaching, bioturbation, or early diagenesis of carbonate sediments; usually preserved by cement filling.

blanket: extensive deposit of reduced thickness.

blow-out: erosional channel form produced by wind.

bottomset: horizontal set of beds or laminae at the toe of foresets.

boudinage: deformation due to stretching and shearing of competent beds, induced by extensional tectonics, or detachment of slide masses (slumping); *syn.:* pull-apart structure.

Bouma sequence: vertical sequence of structures in a turbidite layer; a complete sequence consists of five intervals, or divisions (a through e).

bounce cast: a type of tool mark.

boundary shear stress: tangential pressure transmitted by a fluid to the interface with a solid or other fluid.

brush cast: a type of tool mark.

burrow mottling: patchy structure due to burrowing (bioturbation).

cabbage-leaf casts: *see* frondescent marks (plate 97).

calcrete: calcareous crust in soil of arid-semiarid zones; *see* duricrust.

calc-sinter: speleal encrustation, i.e., stalactite, stalagmite.

caliche: synonymous with calcrete.

candle-end flute casts: morphological variety of flute casts.

carbonaceous: rich in vegetal organic matter.

carrying capacity: capacity of sediment transport of a current.

cascade folding: folding style common in slumps and slides.

channel deposits: sediments confined in a stable or migrating channel; coarse materials represent bars formed when the channel is active; finer materials plug the channel after its abandonment.

channel fill: broadly speaking, synonymous with channel deposits; in a more strict sense, materials deposited after the channel has been disactivated.

channel-form: channel-like profile preserved in stratigraphic sections.

channelized deposits: *see* channel deposits, channel fill.

channel-lag deposit: coarse materials paving the bottom of a channel.

chaotic deposits: breccia, debris flow deposits, slumped beds.

charcoal lineation: streaks and laminae of carbonized vegetal debris.

chatter marks: scratches or cracks made on a rock surface by a mass moving over it; can be either a sedimentary structure (made by glaciers or debris masses) or a tectonic structure on a fault plane.

chevron fold: also called zigzag fold; a tectonic structure.

chord: distance between adjacent crests of ripples, dunes, or other bed forms; equivalent term: (wave) length.

chott: coastal lake, sabkha (north African term).

chute: minor channel incising a fluvial bar.

chute-and-pool structure: erosional bed form of upper flow regime; similar to truncated dunoids.

clast: fragment of rock or skeleton; *see* plates 38, 39.

clastic dike: sedimentary dike constituted by clastic sediment.

clastic sill: concordant injection of clastic sediment.

clast supported: clastic fabric in which framework grains are in mutual contact.

clay ball: clay pebble, globular chip.

clay chip: intraformational clast made of clay or mud; *cf.* intraclast.

clay dune: eolian dune made of clay fragments.

climbing surfaces: surfaces connecting the stoss sides of climbing ripples; a type of false bedding.

clinostratification: inclined stratification (bedding), diagonal stratification (bedding); includes *frontal,* or progradational types (foreset bedding), and *lateral,* or transversal types (epsilon cross-bedding). The term should be used only for large bodies (vertical scale exceeding 1 meter), otherwise it is more properly defined as tabular cross-bedding.

coarsening up: upward increase in grain size in bed*sets* and sedimentary bodies, *not* in individual beds (where the term *inverse grading* should be used).

coastal drifting: long shore transport.

coastal sabkha (sebkha): arid-land coastal plain encrusted by salt, with diagenetic (vadose) precipitation of evaporites.

coated grains: particles surrounded by one or more envelopes of biological, chemical, or biochemical origin; they include oncolites, rhodoliths, ooids, oolites, pisoids, pisoliths.

collapse breccia: fragments due to roof falling in a cavity created by dissolution of buried beds; *syn.:* karst breccia; solution breccia.

comet-shaped flute casts: morphological variety of flute markings.

compaction: consolidation of sediments under load, with loss of fluids and pore space.

competence: the term can have a hydraulic or rheological meaning; in the first respect, it indicates the maximum velocity of a current (related to the size of transported particles); in the second, it is equivalent to strength or resistance to deformation. A *competent* bed, or material, can have a ductile, a brittle, or a mixed behavior.

complex ripple marks: *see* interference ripple marks.

cone-in-cone: load structure consisting of stacked cones pointing downward; it is found in mudstones, shales, and coal. The conical surfaces are surfaces of shear.

contourite: deposit of a contour current (deep-water thermoaline or geostrophic circulation).

coquina: shell bed.

corkscrew flute casts: morphological variety of flute casts.

corrasion ripples: structures similar in shape to ripples on rock surfaces, caused by mechanical (tectonic, glacial) or chemical action.

coset: group of sets of cross beds or cross laminae.

crawling trails: traces made by reptating organisms.

creep, creeping: slow mass movement of particulate material controlled by gravity on slopes.

crescentic ripple marks: lunate ripple marks.

crest lines: lines separating lee side from stoss side in ripples, dunes, and other bed forms.

crevasse splay: apron or lobe of sediment got out of a channel through a breach (crevasse).

crinkles: small or microscale folds.

critical slope: *see* angle of repose.

cross-grooves: groove marks that intersect each other (usually at angles less than 40°).

crumpled ball: ball-and-pillow structure, slump ball.

curled mud flakes: mud curls, curled mud polygons; structures related to desiccation.

current crescents: crescent marks.

current lamination: lamination produced by a tractive current.

current lineation: alignment of grains or other objects parallel to a current.

cuspate ripple marks: *see under* linguoid.

cut-off: erosional event that leaves a segment of a fluvial channel isolated from the main course.

cycloid: corresponds to ball-and-pillow structure, but has been interpreted as liquefaction structure and attributed to earthquake shocks.

datum: correlated stratigraphic level based on physical (lithological, geometric) evidence or data interpretation (biostratigraphic, magnetostratigraphic, etc.).

debris flow: gravity flow where fine particles and water form a single phase (mud), which can support coarser, usually heterogeneous materials; *see also* mud flow.

debris line: landward boundary of storm waves, underlined by debris of various origin.

deep-sea fan: submarine fan, built by turbidity currents and other mass flows.

deformed cross-bedding: deformations include over steepened foresets, folded foresets (gravity failure) and "hooked" tops (current drag).

dendritic rill marks: morphological variety of rill marks; *see* plate 85 and color photo 25.

depocenter: area of maximum accumulation within a basin.

depositional interface: upper surface of newly deposited sediment; ground surface in subaerial conditions, bed in rivers, bottom in subaqueous environments.

depositional strike: direction of contour lines on a slope; direction perpendicular to flow vectors in gravity controlled movements.

desiccation breccia: layer of mud completely broken by subaerial cracking; *see* color photo 27.

desiccation polygons: polygonal mud crusts delimited by desiccation cracks; *syn.:* mud crack polygons.

detachment point: *see* separation point (hydraulic concept).

detachment surface: surface from which a slide or slump detaches; *see* slide scar, slump scar.

dewatering: more or less violent expulsion of water from a compacting sediment.

diagenetic environment: physical and chemical conditions of buried sediments, especially localized in pores and vugs.

diagonal bedding: *see* clinostratification; the term should not be used for ordinary cross-bedding.

diamictite: mudstone with matrix-supported ruditic clasts (angular or rounded, indifferently), in all possible proportions. When pebbles are dominant, the term *pebbly mudstone* is also used.

diamicton: unconsolidated sediment with admixture of fine and coarse materials.

diapiric structures: domes and other structures related to salt injection in overlying beds (and possibly, surface outpouring and flowage).

diastem, diastemic surface: interruption of sedimentation, surface of stratigraphic discontinuity. It is an old term, coined when sedimentation was conceived as a continuous process, to stress gaps. Now, all bedding surfaces are regarded as diastems.

dip: orientation (immersion) of an inclined plane; can be primary (depositional) or secondary (tectonic).

disorganized: attribute of a ruditic sediment or rock, when there is no order in the arrangement of clasts; *syn:* chaotic.

distal: far from the source or the entry point (*see*) of clastics.

domichnia: ecological type of trace fossil, or ichnofossil, left by organisms in living position; *see* dwelling burrow.

downlap: basal contact of a set of inclined beds on a horizontal or less-inclined surface.

downward shift: seaward displacement of the shoreline, regression.

draa: eolian dune of large size and complex structure.

drag fold: minor fold due to relative movements of competent and incompetent beds.

drag mark: mark left by an object dragged at the base of a flow; *syn:* groove cast, chevron cast.

drag wrinkles: small-scale soft-sediment deformation produced on a surface by friction with a moving mass.

drape: sediment cover of a rough substratum; mud cover on a coarser sediment.

dripstone: calcareous encrustation formed by dripping water in caves; includes stalactite and stalagmite.

dropstone: coarse particle dropped by rafting bodies (icebergs, trees, etc.) on subaqueous sediments.

dune bedding: cross bedding at a scale of a dune. The term is imprecise, because subaqueous dunes have an arbitrary upper limit of 1 m in height, whereas there is no limit for eolian dunes.

dunelike form: *see* dunoid.

dune-phase: range of hydraulic conditions allowing the formation of dunes on the bottom.

dunoid: dunelike bed form common in pyroclastic deposits. The profile is either symmetrical or asymmetrical, the internal laminae dip down flow but with an angle smaller than that of repose. The form can be truncated by erosion on the back side.

dwelling burrow: ecological type of trace fossil; *see also* domichnia.

edgewise (conglomerate, coquina): subvertical disposition of coarse particles in ruditic sediments and shell beds.

effective pressure (load): vertical pressure exerted by a submerged sediment, where only the weight of solid particles is effective (water pressure is neutral); represents the active force of compaction.

elutriation: mechanical selection, or particle sorting, made by an ascending fluid.

elutriation pipe: vertical structure in pyroclastic deposits, where fine ash particles have been removed by ascending gas.

entrainment threshold: critical value of shear pressure transmitted by a fluid to remove a stationary particle.

entry point: point source of sediments along a basin margin (shoreline or base of slope).

eolianite: cemented eolian (dune) deposit.

episodic: sporadic; in sedimentology, can be synonymous with catastrophic.

epsilon (cross)bedding: diagonal bedding caused by lateral (across-flow) accretion.

escape burrow: trace left by an animal trying to keep pace with rapid sedimentation.

esker: localized sedimentary body accumulated by melting water in a glaciated area.

evaporite: sedimentary rock constituted by salts precipitated from a solution because of evaporation of water.

evaporite solution breccia: collapse breccia due to the subsurface dissolution of evaporites; can remain as the only witness to vanished evaporites.

external structures: structures present on bedding surfaces; also interfacial structures

extrabasinal: deriving from outside the basin; *cf.* allochthonous, intrabasinal, terrigenous.

fabric: spatial arrangement of sedimentary particles (the term is used as well in igneous and metamorphic rocks). Sedimentary fabric is related to sedimentary processes (*ex.*: imbrication).

faceted pebble: ventifact, pebble shaped by wind.

facies: 1) lithological, textural, structural, and geometrical characters of a generic deposit (descriptive concept); 2) the same features as indicators of a *type* of process or environment (genetic, process-oriented concept); 3) features of a specific sedimentation event or episode (genetic concept, space and time dependent).

facies analysis: sedimentological analysis carried out in the field, based on macroscopic characters of sediments.

facies association: genetically and spatially related facies

forming a sedimentary body or part of it, and reflecting a sedimentary environment or subenvironment.

facies tract: lateral sequence of genetically related facies.

fair weather deposits: beach sediments whose characteristics are controlled by normal waves; opposed to storm deposits.

fall velocity: important parameter for defining the hydraulic size of sedimentary particles; not to be confused with the velocity of sedimentation, or accumulation rate.

false bedding: pseudo stratification simulated by tectonic joints, cleavage, schistosity or sedimentary structures; *see also* climbing surfaces.

fan delta: coarse-grained delta fed by a torrential stream.

fanglomerate: coarse, poorly sorted conglomerate deposited in an alluvial fan or fan delta.

fan-shaped flute casts: morphological variety of flute casts.

feather marking: syn. of frondescent casts.

feeding burrow: ecological type of trace fossil.

feeding trail: ecological type of trace fossil (on bedding surfaces).

festoon: syn. of trough, the basic unit of concave crossbedding; festoon cross-bedding = trough crossbedding.

fining up: upward decrease of grain size in a sequence, or set of beds (*not* in individual beds, where the term *normal grading* should be used).

fissile: attribute of flaggy or laminated rocks with numerous parting planes.

flame structure: pointed pelitic injection; associated to load structures.

flaser bedding: thin, concave-up mud(stone) lenses within sand(stone).

flash flood: instantaneous flood surge in ephemeral streams of semiarid regions.

flat festoons: see swale, swaley cross-bedding.

flat-pebble conglomerate: intraformational conglomerate whose clasts derive from thin crusts or brittle, early indurated sediments.

flat-topped ripple marks: modified ripples with truncated crests.

flint: chert, crystalline silica.

flow regime: hydraulic condition defined by the Froude number, approximated by the ratio between velocity and depth (thickness) of a current, e.g., lower (subcritical) and upper (supercritical) flow regimes.

flow roll: syn. of pseudonodule, ball-and-pillow, slump ball.

flow separation: detachment of flow lines and fluid shear from the boundary; *cf.* attachment point, separation point.

flowstone: calcareous deposit encrusting cave walls or paving cave floors, caused by water flowing along them; *cf.* dripstone.

fluidal structures: generic term for whirling traces in deformed sediments.

fluidization: mixing of gas and solid particles that creates a phase behaving like a liquid; process occurring in pyroclastic flows.

foam marks: delicate structures (wrinkles) in littoral sands; similar to current markings on lamina surfaces of turbidites.

foredune: sand mound behind grass tufts or other obstacles in backshore zones.

foreset bedding: basinward dipping beds accumulating on a submerged slope.

foreshore: intertidal zone of a beach; its upper part is the swash zone.

form set: set of laminae recording a bed form in section.

founder breccia: *see* collapse breccia.

fragmented bed: bed with a chaotic, or disorganized, internal structure.

freezing: consolidation of a moving sediment-fluid mixture.

frost cracks: structures related to cryoturbation.

furrow: elongate erosional mark; *see* ridge-and-furrow structure.

gas hole: surface expression of escaping gas; on a large-scale: pockmark.

gas pit: *see* gas hole.

Gilbert delta: fan delta advancing in a relatively deep body of water. The resulting geometry shows a typical "triad": topset, foreset, and bottomset beds (*see* color photo 2); *see* clinostratification, fan delta, progradation.

glacial flutes: marks made by glaciers on bedrock, similar to current-produced flutes in sediments.

glacial striation: drag structures made by glaciers on both sediment and rock.

glaciotectonic structures: soft-sediment deformation caused by overriding ice.

glide breccia: breccia caused by sliding or tectonic friction.

graded bedding: fabric expressing *vertical* grading.

graded interval (division): basal part of the Bouma sequence.

grading: vertical or areal variation of grain size in beds, layers, and laminae, i.e., in *individual* depositional units; can be normal or reverse (inverted).

grain fall: gravity-driven deposition of individual particles at the foot of slopes and cliffs; *cf.* rockfall.

grain flow: collective movement of solid particles behaving like a fluid.

gravel stripe: alignment of pebbles parallel to current or wave motion.

grazing mark: scratch made by an animal on the bottom surface; trace fossil related to feeding or crawling activities.

ground surge: type of pyroclastic flow, observed for the first time at the base of ascending columns in experiments with atomic bombs.

growth fault: normal fault induced by gravity on the margin of a basin (downthrown side is basinward).

growth model: mode or style of accretion of a sedimentary body; examples are aggradation, progradation, frontal accretion, lateral accretion.

gully: minor channel on a slope.

hailstone imprints: small pits similar to raindrop imprints, preserved in fine sediments; also (syn.) pitted mud.

hemipelagic: attribute of fine sediments composed partly of indigenous (intrabasinal) particles, like microfossil tests, and for the other part by terrigenous (land derived) or extraformational (from older geologic bodies) materials.

herringbone: attribute of a variety of cross-bedding.

heterolithic: attribute of a facies showing more than one lithology; virtually restricted to sand(stone)/mud(stone) alternations; *cf.* rhythmic bedding, rhythmite.

hexagonal cross-ripple marks: *see* interference ripple marks, tadpole nests.

hiatus: discontinuity of sedimentation, stratigraphic gap.

hieroglyph: old name for sole marking.

horizontal lamination: the term should be used only in Modern and Recent sediments. In Anciet sediments, use plane-parallel lamination, bed-parallel flat lamination.

host sediment: sediment embedding secondary structures; *see* diagenetic environment.

hydraulic jump: local hydraulic instability of a current owing to sudden increase in thickness and/or change in gradient.

hydraulic size: fall velocity of a sedimentary particle, controlled by size, weight and shape.

hydraulic sorting: grain selection based on hydraulic size.

hydroplastic deformation: ductile deformation in water-soaked sediments..

hydrostatic pressure (head, load): pressure exerted by water (in all directions).

hypersaline: marine water with salt concentration above normal.

hyposaline: marine water with salt concentration below normal; *syn.:* brackish.

ice-crystal marks: indicators of cryoturbation.

ice-push structures: *see* glaciotectonic structures.

ice-rafted sediment: includes dropstones.

ice ripples: undulations on ice surface due to contact with water.

ice-shear structures: *see* ice-push structures.

ice-wedge polygons: polygonal zones of a ground or soil delimited by ice wedges.

ice wedges: ice-filled cracks in frosted soil or sediment.

ichnofossil: trace fossil, organic structure modifying a sediment.

imbrication: shingled arrangement of platy particles, produced by currents or waves.

impact marks: traces left by impacting objects on sediments.

inclined bedding: *see* clinostratification. These terms imply depositional slopes; for inclination of tectonic origin, use *tilting*.

incompetent: easily deformable material.

indicator: sedimentary character that records a specific mechanism, process or environment.

infauna: benthic animals living inside the sediment.

injection structures: deformation structures related to upward intrusion of various materials: water (*see* dewatering), quicksand, fluid mud, plastic mud, salt; also sedimentary dikes, sedimentary sills, diapiric structures, liquefaction structures, fluidization structures, dish structure, pillar structure, and others.

inland sabkha (sebkha): subbottom of a dried lake, with salt encrustations; *cf.* playa lake.

interfacial structures: marks, imprints on bedding planes.

interference ripple marks: a rippled surface with different orientations of ripple crests.

internal sediment: particles that infiltrate in a porous sediment or a leaching cavity.

intertidal: comprised between average high tide mark and low tide mark.

intertonguing: interfingering of depositional units (regardless of scale).

intrabasinal: relates to materials and phenomena originating within a sedimentary basin; *cf.* allochthonous, extrabasinal, terrigenous.

intraclast: clastic particle deriving from fragmentation of intrabasinal sediment.

intraformational: deriving from intrabasinal reworking.

intrastratal: internal to beds or layers. Examples: intrastratal solution, intrastratal structure.

inverse grading: upward increase of grain size within a bed or layer; *cf.* coarsening-up.

jet: expanding flow.

joint: fracture, surface of tectonic discontinuity without displacement of parts.

jökulhlaup: Icelandic term for catastrophic flood deriving from the collapse of an ice dam.

kames: morphological and sedimentary unit deriving from deposition by melting water in glaciated areas; *cf.* esker.

kankar: calcareous duricrust.

karst: subterranean cavities in soluble rocks; also cave deposits, dripstone, flowstone, speleothems.

keeled ripple marks: morphological variety of r.m.

key bed (layer): bed (layer) used as a marker for stratigraphic correlation; corresponds to a time line or synchronous stratigraphic level.

kinetic sieving: downward shift of fine particles between larger ones in a moving sediment.

lag (conglomerate, deposit, pavement): coarse particles lagging behind most currents which are not able to carry them. More or less continuous cover of erosional surfaces (except cut-and-fill scours).

lahar: dry or wet avalanching (debris flow) of *remobilized* volcanic ash, often embedding coarse particles; *cf.* pyroclastic flow.

laminar sublayer: part of a boundary layer in close contact with the boundary, where the flow is laminar (not turbulent).

laminaset: packet or bundle of conformable laminae.

lateral growth: mode of accretion of a sedimentary body.

layer-cake: a growth model, with vertically accreting, parallel, continuous layers.

leaching vug: *see* birds-eye structure.

lebensspuren: German term for trace fossils.

lee side: frontal or anterior side of a bed form.

lensing: thinning at both sides of a bed or a body.

lenticular bedding: alignment of sand(stone) lenses embedded in mud(stone).

levee: natural or artificial super elevation of a channel bank. Natural levees are built by overflowing sediment.

limnic: lacustrine.

linear structures: structures reflecting the trend, but not necessarily the direction of a movement.

lineation: linear structure parallel to movement.

linguoid (bar, dune, ripple): morphological variety of a bed form; *syn.:* cuspate.

liquefaction: process destroying the solid framework of a water-saturated sediment and creating a mixture with the behavior of a liquid.

lithosome: rock body.

lithostatic pressure (load): vertical pressure exerted by rock bodies, including the solid particles of sediments.

lithotope: part (area) of a sedimentary environment characterized by a certain lithology.

load: static pressure due to weight.

load-casted (flute cast, groove cast, etc.); sole markings emphasized by load effects.

longitudinal: parallel to a movement or to the length of an object; e.g., longitudinal bar, longitudinal section.

longitudinal furrows and ridges: type of wrinkle marks; *see* ridge-and-furrow structure.

long shore bar: submerged littoral bar, parallel to shoreline.

long shore current: current parallel to shoreline, developing in beaches where and when waves impinge obliquely.

low-angle(cross-bedding, cross-lamination): inclination smaller than the angle of repose in laminae or beds.

lower flow regime: *see* flow regime.

low-flow structures: structures associated to bed forms of lower flow regime.

lumpy bedding: *see* mottling, nodular bedding.

lunate (ripples, dunes): morphological variety of bed forms; *syn.:* barchanlike, barchanoid.

lunette: *see* clay dune.

maar: *see* tuff ring.

marker: correlatable stratigraphic horizon, recognizable in the field; *cf.* datum; key bed.

mass flow: flow transporting high quantities of sediment per unit area and unit time. It can be fluid or gravity driven.

massive: ambiguous attribute, used both for thick or very thick beds and for the lack of structures (the two characters often coincide but should be kept separate).

matrix: finer material in sediments with two size *modes,* filling pore spaces or embedding coarser particles.

matrix-supported: surrounded, embedded by matrix.

meander: loop of a migrating channel with high sinuosity.

meander scroll: *see* scroll bar.

megaripple: long ripple or subaqueous dune.

micrite: microcrystalline calcite deriving from crystallization of carbonate mud; opaque in thin section, whitish in outcrop and polished section.

microdelta: modification structure observed in intertidal environments and looking like a small-scale delta; *see* plate 29.

mima mounds: patterned ground with moundlike relieves made of silt; attributed to cryoturbation or earthquakes; occur mostly in cold regions.

mottling: patches of different color or texture due to bioturbation or diagenetic reactions (nodule growth, etc.).

mud ball: *see* clay ball, mud pebble.

mud clast: intraclast, clay chip.

mud crack: desiccation crack.

mud-draped ripples: ripples with tops preserved by a mud cover; *see* plate 58.

mud flat: mud-covered part of a tidal or coastal flat.

mud flow: end member of debris flows (no clasts).

mud lump: top of a mud diapir.

mud mound: micritic mass accumulated by organisms.

mud pebble: rounded mud clast; *see also* clay ball.

mud pellet: *see* mud pebble.

mud ripples: undulations on mud surfaces subject to current or wind shear; a rare structure.

mudstone: fine-grained rock, or compacted sediment, with no obvious fissility; textural term, independent from composition.

mudstone conglomerate: *see* intraformational conglomerate.

mudstone pebbles: rounded intraformational clasts.

multiple grading: repeated grading; can occur in laminae within beds or in amalgamated beds within a bedset.

nodular bedding: pervasive nodular structure that destroys original bedding; *see* plate 171.

nodule: structure of chemical or biochemical origin; small mass of amorphous or crystalline mineral growing within a host sediment.

normal grading: upward decreasing size in a clastic bed or layer; *cf.* fining-up.

obstacle shadow: sediment accumulation behind an obstacle.

offshore bar: sedimentary relief on a shelf, separated from the littoral zone.

olist(h)ostrome: slid mass or big debris flow deposit (not clear in the original definition).

oncoid: *see* oncolith.

oncolith: grain coated by calcified microbal laminae.

oncolitic structure: concentric structure of oncoliths.

onlap: a type of discordant contact; beds leaning against an inclined surface.

organic mound: sedimentary relief caused by building organisms; e.g., reef, bank, mud mound, stromatolitic mound.

organized: attribute of a coarse clastic deposit, in which the particles are arranged with some order (layering, size or shape sorting, imbrication).

oscillation ripples: wave ripples.

outwash deposits: glacio-fluvial sediments.

overbank deposits: fine sediments overflowed from channels.

overfold: overturned fold, often occurring as a broken, uprooted hinge in slides.

oxbow lake: abandoned meander loop.

packing: "geometric" density of grains; ratio between volume of framework grains and total sediment volume.

parabolic dune: type of coastal dune with slipface convex downwind (opposite to barchan dune).

paralic: attribute of coastal to shallow water environments, or facies; *syn.:* marginal marine.

parallel lamination: plane or wavy, continuous lamination.

parting plane lineation: alignment of grains or objects on an intrastratal surface (parting plane).

parting-step lineation: parting lineation with shallow terraces.

pebbly mudstone: type of diamictite. The origin can be glacial (till), glacio-marine or glacio-lacustrine (dropstones) or related to mass flows.

penecontemporaneous: early post-depositional.

penecontemporaneous deformation: *see* soft-sediment deformation.

permafrost: permanently frosted ground.

phacoidal: peculiar shape of intraformational fragments.

phreatomagmatic: qualifies an explosive eruption involving water bodies.

pillar structure: small-scale sedimentary dike within sandstone beds, commonly associated with dish structure.

pinch-out: lateral termination (wedging) of a bed or sedimentary body.

pit-and-mound structure: *see* gas hole, gas pit.

planar cross-bedding: geometrical variety of cross-bedding, contrasted with trough (concave, festoon) cross-bedding; *syn:* tabular cross-bedding.

plane bed: flat depositional interface, with no fluid movement over it or a current in upper flow regime.

plane bed phase: range of hydraulic conditions preventing formation of bed forms.

platform: environment of shallow water carbonate deposition.

playa lake: ephemeral lake of semiarid regions; *cf.* inland sabkha.

point bar: laterally accreting bar in a meander loop; *cf.* epsilon (cross)bedding.

polygonal ground: patterned ground of subpolar regions characterized by alternating freezing and thawing; *cf.* ice-wedge polygons.

preservation potential: determines the absolute and relative abundance of structures that are recorded in sedimentary rocks, with respect to the total original number.

priel: erosional structure at the base of beds; examples: scour, gutter cast.

primary lineation: *see* parting plane lineation.

prismatic cracks: variety of desiccation cracks.

problematic markings: sole marks of unexplained or not completely clear origin.

prod marks: morphological variety of impact markings.

progradation: a growth model of sedimentary bodies; lateral (frontal) accretion in a water body. Frontal means "in the direction of main flow vector and sediment supply," in contrast with transversal, or normal to flow vector; *cf.* clinostratification, foreset bedding, Gilbert-type delta.

prograding: advancing basinward.

progressive cross-bedding: syn. of climbing ripple cross-lamination.

proximal: close to the source or the entry point (*see*) of clastics.

proximal rim: upcurrent margin of a structure.

pseudo bedding: *see* false bedding.

pseudo breccia: old term for breccias of nondepositional origin.

pull-apart (basin): basin originated by wrench, or transcurrent faulting.

pull-apart (structure): extensional deformation; *see* boudinage.

pyramidal dune: a morphological variety of eolian dune; *syn.:* star dune.

pyroclastic flow: broadly speaking, any dispersion (solid-fluid mixture) of volcanic ash and hot gases flowing under the control of gravity; strictly speaking, a laminar nonturbulent flow with a high concentration of solid particles, similar to a debris flow.

pyroclastic surge (flow): highly turbulent pyroclastic flow with tractive power (*cf.* dunelike bed forms). Base surge and ground surge are two varieties.

quickclay: liquefied clay.

quicksand: liquefied sand.

radial: type of fabric or growth pattern in chemical and biogenic structures.

rafting: transport by floating objects.

reactivation surface: small-scale unconformity between foreset laminae.

reattachment point: *see* attachment point.

recumbent fold: overturned fold with horizontal or slightly dipping axial plane.

recurrence time: return time of an event.

recycled sediment: sediment deriving from erosion of an older deposit or sedimentary rock.

recycling: removing particles from an older sediment, extraformational reworking.

redeposition: deposition of sediment removed from a previous site of accumulation within the basin perimeter; *cf.* intraformational reworking, intrabasinal reworking.

resedimentation: redeposition caused by mass transport and catastrophic events.

residual angle (after shearing): slope angle assumed by a loose sediment after avalanching took place; *cf.* angle of repose.

resting tracks: ecological type of trace fossils, due to stationing animals.

reversed density gradient: cause of gravitational instability in stratified fluids or sediments; produces structures such as load casts, pseudonodules, flame structure, convolutions.

reworking: in paleontological jargon, remobilization of body fossils; in sedimentological terms, any action of mechanical or biological *modification* of a previous sediment.

rheotactic fabric: particles reoriented by a current after deposition.

rhodolith: calcareous nodular body produced by algal accretion around a nucleus.

rhomboid rill marks: morphological variety of rill marks; *see* plate 84.

rhomboid ripple marks: lozenge-shaped ripple marks.

rhythmic (bedding, lamination): regular repetition of a couplet, or a triplet, of lithologies.

rhythmite: rhythmically bedding or laminated deposit; the term is employed mostly for thin beds or laminae (e.g., varve).

ridge-and-furrow structure: variety of longitudinal sole marks, produced by "spiral tubes" (*see*); preserved both as original (on parting surfaces) and as a mold.

riffle-and-pool: channel bed morphology in sinuous streams.

rill molds: variety of flute casts, mistakenly interpreted as molds of rill marks.

rim cement: crystalline coating of particle surfaces or cavity walls; passes into void or other type of (central) cement; present in geode, druse, hollow concretions.

rip current: return (seaward directed) current active after storms in littoral zones; it is laterally confined and can scour channels.

ripple crest: limit between the two oppositely dipping sides of a ripple (not necessarily the top).

ripple fan: set of "parasite" ripples flooring the trough before a dune; *see* plate 29.

ripple index: describes the basic shape of a ripple by the length/height ratio.

ripple-load convolutions: convolutions related to load-casting of rippled sand beds.

ripple trough: the deepest part of a ripple.

rock avalanche: dry or wet sliding of rock debris; a type of very coarse sediment gravity flow; *syn.: sturzström* (Swiss term).

rockfall: downslope fall of fragments detaching from a rock mass (or an organic reef). Particle by particle gravity accumulation.

rockslide: type of landslide; mass of rock sliding as a single block or in fragments (rock avalanche).

rollability: propensity of sedimentary particles to roll.

roll casts: marks made by rolling objects; a type of tool marks.

rootlet bed: bed with fossil roots or root traces, often underlying a coal bed.

roughness: asperities of a bed or solid boundary; in hydraulics, it is not the absolute relief that counts, but its relation with the flow depth.

roundness index: measures the amount of abrasion on edges and corners of clastic particles.

ruffled groove cast: groove cast with "barbs" on one or both sides, making transition to chevron cast; *see* plate 101.

runnel cast: fill of an elongated scour at the base of a bed; *see* gutter cast.

runzelmarken: German term for wrinkle marks.

sabkha: salt flat with vadose evaporites; *see also* coastal sabkha, inland sabkha.

saltation: mode of bed load movement, with grains lifted temporarily in the flow; particularly diffused in wind transport.

sandblasting: a type of abrasion in eolian environments.

sand blow: dewatering or liquefaction structure; *see* pillar structure, sand volcano.

sand boil: *see* sand blow.

sand flat: sand covered part of a tidal flat; *see* plate 28.

sand flow: variety of grain flow, with sand moving like a fluid substance.

sand holes: gas holes on a sand bed; *see* plate 149.

sand pipe: *see* pillar structure.

sand ribbon: thin, elongated sand body deposited by wind in subaerial environment or tidal currents in shallow submarine environment.

sand shadow: sand accumulated behind an obstacle; *cf.* foredune; *see* color plate 12.

sand slough: *see* sand blow.

sand spout: *see* sand blow.

sandstone whirlball: a spiraled variety of pseudonodule (pillow) or slump ball.

sandur: glacio-fluvial plain.

sand wave: large scale, subaqueous, migrating bed form made of sand(stone), produced by a tractive current.

sausage structure: *see* boudinage.

scallop: dissolution furrow, *karren.*

scour-and-fill: scour immediately filled by sediment carried by the same scouring agent.

scour marking: erosional mark carved by current eddies; includes flute mark, crescent mark, gutter cast, current crescent, moat, etc.

scour remnant ridges: "passive" relieves contrasted with "active" ones caused by deposition or deformation.

scroll bar: minor arcuate bar on top of a point bar.

sedentary benthos: benthic organisms, individual or colonial, which remain fixed to the bottom.

sedimentary interface: *see* depositional interface.

sedimentation event: single act of deposition.

sedimentation episode: sequence of events related to the same process or environment.

sedimentation unit: the record of a sedimentation event, or episode; lamina, bed or layer, laminaset, bedset.

sediment flow: *see* debris flow.

sediment wave: large scale undulation of sea bottom due to a current; seismic reflectors reveal unidirectional migration.

seepage: infiltration and percolation of fluids in the sediment pores.

seif: longitudinal desert dune.

separated flow: fluid shear transferred from the bottom to the fluid body.

separation bubble: shadow zone on the down current side of a bottom asperity (roughness element), where a secondary circulation separates from the main flow.

separation point: point of detachment of a separated flow, upcurrent limit of a separation bubble; *cf.* (re)attachment point.

separation zone: *see* separation bubble.

shaling out: lateral transition into shales.

shear (force, stress, strain): reciprocal mechanical action exerted by masses in relative movement parallel to their surface of separation (tangential).

shear plane: *see* shear surface.

shear surface: surface of separation of masses sliding one past the other.

shear strength: resistance to tangential movements (sliding).

sheet: shape of a layer, or sedimentary body whose areal extent is much prevalent on thickness.

sheet crack: desiccation crack parallel to bedding.

sheet erosion: instantaneous erosion caused by an unconfined current.

sheet flood: flooding current not confined to a channel.

sheet flow: unconfined flow, with limited depth relative to width.

shelly bed: accumulation of organic (skeletal) remains.

shingle structure: *see* imbrication, imbricated fabric.

shoreface: permanently submerged part of a beach, beneath low tide mark.

shrinkage crack: subaqueous crack caused by physiochemical contraction in a mud.

silcrete: siliceous duricrust.

sinter: chemical encrustation in springs, lakes, and streams; can be calcareous (travertine, encrusting limestone) or siliceous.

skeletal: attribute of hard, mineralized parts of organisms.

skim mark: syn. of chevron mark.

skip cast: variety of impact tool casts.

slide scar: bare surface left by the detachment of a slide.

slide sheet: tabular accumulation of a slid mass.

slipface: down current side of a bed form, where grain avalanching takes place.

slip plane (surface): surface of sliding, base of a sliding mass; *cf.* shear plane.

slump: rotational sliding.

slump ball: pillow structure entrained in slumping or sliding.

slump fold: folding related to gravity sliding.

slumping: *see* slump.

slump overfold: overturned slump fold.

slump scar: curve, spoon-shaped surface of detachment; *see also* slide scar.

slump sheet: tabular accumulation of a slumped mass.

small-scale cross-stratification (bedding): cross-lamination, ripple cross-lamination.

soft-sediment deformation: penecontemporaneous deformation, deformation occurring in the sedimentary environment or basin.

solifluction: soil creep.

solution furrow: surficial karstic feature; *syn.*: scallop, karren.

solution ripples: surface karstsic feature; pseudo ripples caused by dissolution

sorting: selective distribution of grains, based on size, shape, or weight.

sparite: calcite in large, clean crystals.

spasmodic event: catastrophic or episodic event, suddenly releasing a high amount of energy.

speleal: related to caves, karstic.

speleothem: cave deposit.

spiral tubes: longitudinal vortexes, where fluid particles follow helicoidal paths.

splitting planes: parting planes of fissile rocks.

spongy (texture, rock): vuggy.

spontaneous liquefaction: *see* liquefaction.

spreiten (Ger.): minor biogenic structures associated with trace fossils.

spring pits: small structures produced on the surface of beach sand by escaping water.

spring-neap cycle: tidal cycle corresponding to half a month (lunar month = 27.3 days).

spur: accessory structure of ripple marks, parallel to the flow.

spur-and-groove: morphology of the submerged rim of a reef.

squamiform load casts: flattened load casts; *cf.* load-casted current markings, syndromous load casts; *see* plates 124, 125 **A.**

squeezed mud ridges: *see* flame structure.

stalactiform: looking like a stalactite.

standing wave: stationary undulation present on the surface of both water and water-sediment surface in the antidune phase (upper flow regime) of a tractive current. The sediment is in motion on the bottom.

star dune: syn. of pyramidal dune; type of desert dune.

starved basin: undernourished basin, basin with little input and/or production of sediment.

starved ripples: ripples forming a discontinuous sand bed (the troughs are occupied by mud).

steady flow: a flow whose configuration (expressed by the velocity profile) does not change with time.

stone cluster: *see* pebble cluster (to be used for clasts larger than pebbles).

stoss side: upcurrent side of a bed form.

straight crested ripples: contrasted with sinuous and three-dimensional ripples.

streaked-out ripples: old term for flame structure; nothing to do with ripples.

streamline: flow line, connecting velocity vectors in the flow direction.

striae: striations, scratches; delicate markings.

strike: intersection between an inclined plane and the horizontal, e.g., depositional strike, tectonic strike.

stromatolite: laminated structure deriving from lithification of an algal (microbial) mat.

structureless: term recommended instead of *massive*, for beds of uniform or homogeneous texture.

sturzström: Swiss term for rock avalanche.

subcritical flow regime: *see* flow regime.

sun crack: *see* desiccation polygons, mud crack.

superposed ripple marks: *see* interference ripple marks.

surge: rapid uprising of energy, catastrophic wave or flow (in pyroclastic processes); *cf.* base surge, ground surge.

surge (flow): *see* pyroclastic surge.

suspended load: the bulk of particles carried in suspension by a flow.

suspension current: gravity driven (density) current in which the density contrast with surrounding fluid is created by suspended particles; *cf.* turbidity current, pyroclastic surge flow.

swale: 1) furrow between scroll bars of a meander bar; 2) scour related to hummocky cross-bedding.

swaley cross-bedding: a variety of trough cross-bedding with slightly curved concavities; erosional counterpart of hummocky cross-bedding; also known as flat festoon (bedding).

swash: wave dying on the beach face.

swash zone: upper part of the foreshore, delimited by the berm.

symmetrical ripple marks: *see* oscillation ripple marks, wave ripples.

syndromous load-casts: syn. of dendritic ridges-and-furrows.

syneresis: subaqueous shrinkage of gel-like materials.

tabular cross-bedding: planar cross-bedding.

tangential cross-bedding: foreset bedding with asymptotic toe.

tectonic ripples: pseudo ripples; tectonic crenulations simulating sedimentary ripples.

tempestite: storm layer, storm deposit.

tephra: ashfall deposits.

terrigenous: deriving from land areas.

thin-out: *see* pinch-out, wedging.

thixotropy: property of colloidal substances, passing from gel to sol conditions when disturbed.

threadlike flute casts: morphological variety of flute casts.

threshold of grain movement: critical value of fluid stress for removing a grain from the bed; *syn.:* entrainment threshold.

tidal channel: channel in a tidal flat or estuary; *cf.* tidal inlet.

tidal creek: small channel in mud flats.

tidal delta: splay or fan of sand at the mouth of a tidal inlet.

tidal inlet: passage between barrier islands.

till: glacial deposit.

tillite: consolidated glacial deposit.

time-transgressive: diachronous, crossing time lines (surfaces).

toeset: deposit connecting foreset and bottomset beds (laminae).

tool marks: marks made by objects transported by a current, a mass flow, a slide, or a glacier.

toplap: upper contact of foreset beds (laminae).

topset: upper part of the triad top-, fore- and bottom-set; *see* Gilbert delta.

trace fossils: individual bioturbation structures.

track: type of animal trace preserved on bed surfaces.

traction carpet: layer of moving and colliding grains at the base of a flow; collective bedload movement.

tractive movement: selective bedload movement in which particles follow individual paths.

trail: type of animal trace preserved on bed surfaces.

tranquil flow regime: lower flow regime.

transverse ribs: sort of crude ripple-like bed forms built with pebbles.

travertine: concretionary or encrusting limestone due to chemical precipitation in continental environments (lake bottom, river bed, springs). *syn.:* encrusting limestone, calcareous sinter, tufa.

trough: depressed portion of a bed form; longitudinal scour.

trough cross-bedding: morphological type of cross-bedding reflecting the migration of three-dimensional bed forms; *see* festoon, swaley cross-bedding.

truncation: upper termination of depositional units caused by erosion.

tufa: *see* travertine.

tuff ring: circular rampart built by accumulation of pyroclastic products around a vent located in a lake or an area with abundant water in the subsurface; *cf.* phreatomagmatic; *syn.:* maar.

tumble marks: marks made by jumping/rolling objects.

turbidite: the deposit layer of a turbidity current.

turbidity current: a type of subaqueous density current in which the excess density is created by suspended sediment.

turboglyphs: old name for flute casts.

twisted flute cast: morphological variety of sole marking.

twisted groove cast: morphological variety of sole marking.

unconfined current: current not bounded by a channel; *syn.:* sheet flow.

unconformity: geometric discontinuity of sedimentological or stratigraphic nature.

undercutting: erosion at the foot of a cliff or river bank.

undertow: *see* rip current.

upper flow regime: supercritical flow regime; *see* flow regime.

U-shaped burrow: type of shallow-water burrow.

vadose zone: subsurface zone above the water table.

varve: annual or seasonal deposit, commonly rhythmic.

velocity profile: line connecting velocity vectors in a flow section (longitudinal, transversal or horizontal).

ventifact: stone or object sculptured by the wind.

vergence: direction of tectonic transport in zones subject to compression and shortening, which result in folds and thrusts; applied by analogy in soft-sediment deformation caused by tangential stresses.

waning current: current related to a catastrophic event, which looses energy after an initial peak.

wash-out: *see* scour-and-fill.

washover fan: lobate or fan-shaped deposit in a lagoon, caused by a storm breaching a barrier island.

watery slide: *see* debris flow, grain flow.

wave base: maximum depth of traction for waves. Three levels can be distinguished, with reference to fair-weather waves, average storm waves, and exceptional (storm, tsunami) waves, respectively.

wavelength: crest to crest distance in periodical bed forms; *syn.:* chord.

wave-ripple lamination: a morphological type of cross-lamination, reflecting wave ripples.

wave ripples: ripples produced by wave action. Without specification, normal waves are understood (storm waves produce hummocky cross-bedding).

wavy bedding: a member of the triad completed by flaser and lenticular bedding; alternating sand and mud beds are both continuous, with wavy surfaces due to ripples and load effects.

wavy lamination: a variety of parallel lamination; *cf.* plane lamination; contrast with discontinuous (cross) lamination.

way-up criteria: methods for reconstructing the stratigraphic polarity (up and down) in tilted or overturned strata.

weathering: physical and chemical alteration of rocks exposed to the atmosphere.

wedging: lateral thinning of a depositional unit or sedimentary body.

wedge-out: *see* pinch-out.

wedge-shaped cross-bedding: morphological variety of cross-bedding.

wind-drift current: current stimulated by the wind blowing on a water surface.

wind ripples: ripples on dry sand, mostly related to saltation mechanism; *see* ballistic ripples.

wingless flute casts: incomplete flute casts.

winnowing: cleaning of coarse sediment, selective removal of fine particles.

wrinkle marks: delicate erosional or deformational marks, usually preserved on lamina surfaces within beds; type of intrastratal current marks; *see also* longitudinal furrows and ridges.

yield limit: the stress at which a material begins to undergo permanent deformation.

yield strength: *see* yield limit.

younging: going up stratigraphically.

Y-shaped burrow: type of vertical burrow common in shallow water.

zibar dune: a type of desert dune devoid of slipface.

Index